T0176511

Tunnel Field-Effect Transistors (TFET)

Tunnel Field-Effect Transistors (TFET)

Modelling and Simulation

Jagadesh Kumar Mamidala
Indian Institute of Technology (IIT), Delhi, India
Rajat Vishnoi
Indian Institute of Technology (IIT), Delhi, India
Pratyush Pandey
University of Notre Dame, Indiana, USA

WILEY

This edition first published 2017

Registered Office
John Wiley & Sons, Ltd, The Atrium, Southern Gate, Chichester, West Sussex, PO19 8SQ, United Kingdom

For details of our global editorial offices, for customer services and for information about how to apply for permission to reuse the copyright material in this book please see our website at www.wiley.com.

Library of Congress Cataloging-in-Publication data applied for

ISBN: 9781119246299

A catalogue record for this book is available from the British Library.

Set in 10.5/13pt Times by SPi Global, Pondicherry, India
Printed and bound in Malaysia by Vivar Printing Sdn Bhd

10 9 8 7 6 5 4 3 2 1

Contents

Preface

Overview

Just as a human body is made up of millions of biological cells, an integrated circuit is made up of millions of transistors. Transistors are the basic building blocks of all modern electronic gadgets. Ever since the advent of CMOS circuits, the dimensions of the transistor have been continuously scaled down in order to pack more logic on to a silicon wafer and also to reduce power consumption in the circuits. In recent years, with mobile devices becoming popular, the search for low power devices with steep switching characteristics has become important. Highly scaled MOSFETs are rendered unsuitable for low power applications due to a thermal limit on their switching. Hence, the Tunnel Field Effect Transistor (TFET) is being explored extensively for low power applications. A TFET has a steep switching characteristic as it works on the phenomena of band-to-band tunnelling. Over the past few years, TFETs have been heavily researched by various notable groups in the field of semiconductor devices across the globe.

This book provides a comprehensive guide for those who are beginning their study on TFETs and also as a guide for those who wish to design integrated circuits based on TFETs. The book covers the essential physics behind the functioning of the TFETs and also the device modelling of TFETs, for the purpose of circuit design and simulation. It begins with studying the basic principles of quantum mechanics and then builds up to the physics behind the quantum mechanical phenomena of band-to-band tunnelling. This is followed by studying the basic functioning of TFETs and their different structural configurations. After explaining the functioning of TFETs, the book describes different approaches used by researchers for developing the drain current models for TFETs. Finally, to help new researchers in the field of TFETs, the book describes the process of carrying out numerical simulations of TFETs using the TCAD tool Silvaco ATLAS. Numerical simulations are helpful tools for studying the behaviour of any semiconductor device without getting into the complex process of fabrication and characterisation.

Key feature in relation to existing literature

This book is the first comprehensive literature on TFETs, which are very popular transistors and have been extensively studied in recent years; they are going to be important building blocks for low power solid state circuits in the future. It is a one-stop volume for studying TFETs for someone who has a basic knowledge of MOSFET physics. It covers the physics behind the phenomena of tunnelling as well as the device physics of TFETs. It also has a unique feature of describing device simulation along with device physics so as to enable readers to do further research on TFETs.

The presentation of the book is clear and accurate and is written in simple language. The book endeavours to explain different phenomena in the TFETs using simple and logical explanations so as to enable the reader to get a real feel for the functioning of the device. Also, each and every aspect of the TFET has been compared to that of the MOSFET so that the facts presented in the book make more sense to the entire semiconductor device fraternity and help in the integration of the TFET with the prevailing technology in the industry. The book also attempts to cover all the recent research articles published on TFETs so as to make sure that, along with covering the basics, it also covers state of the art work on TFETs.

Key features in relation to existing literature

1

Quantum mechanics

1.1 Introduction to quantum mechanics

Before attempting to investigate the workings of a tunnelling field-effect transistor, it is essential to be familiar with the concept of tunnelling. Tunnelling is a quantum phenomenon, with no counterpart in the everyday physics one encounters, or the physics that one applies while dealing with devices a few hundred nanometres in length. The initial two chapters will, therefore, help us develop an understanding of quantum phenomena. In this chapter, we will present an introduction to the field of quantum mechanics and the next chapter will discuss the phenomenon of tunnelling in detail.

The chapter begins with a description of a landmark experiment that conclusively proved the wave nature of particles, after which we will study the concept of wavefunctions and how to use Schrodinger's equation to obtain them. A few basic problems will be presented so that the readers may familiarise themselves with basic quantum concepts.

1.1.1 The double slit experiment

There are many experiments that led to the conception of quantum mechanics – blackbody radiation, the Stern Gerlach experiment, the photoelectric effect, the line spectrum, etc. However, for our purposes we will concentrate on one of the landmark experiments, that is the double slit experiment, which demonstrated the fundamental quantum nature (i.e. both wave and particle) of electrons.

Tunnel Field-Effect Transistors (TFET): Modelling and Simulation, First Edition. Jagadesh Kumar Mamidala, Rajat Vishnoi and Pratyush Pandey.

You would have read that only waves can undergo superposition, and not particles. Superposition is the fundamental principle behind the occurrence of interference – therefore, if something exhibits interference, it must have a wave nature. The double slit experiment is famously associated with Thomas Young, who used it for the first time in the early nineteenth century to prove the wave nature of light. Before this experiment was performed, light had been associated with a particle nature (since the times of Newton), and the fact that it underwent interference was conclusive proof of its wave nature.

However, the behaviour of light that led Newton and others to believe that it had a particle nature could not be reconciled with this newly formed wave picture. It took another century of research and experiments to establish a rather astonishing result regarding the behaviour of light – that it displays both particle and wave natures. The particle nature leads to phenomena such as the photoelectric effect and rectilinear propagation of light in ray optics; the wave nature explained the interference and diffraction of light.

While this dual nature (that is both particle and wave natures) of light was being worked out, many people were, independently, studying the behaviour of subatomic particles. Phenomenon like the discrete line spectrum of hydrogen, the observed distribution of blackbody radiation, etc., could not be explained by any established theory. Theoretical physicists were in a quandary. At this point, de Broglie hypothesised that, just like light, particles possess a dual nature as well. When de Broglie made this hypothesis, there was little evidence to support his claim. A few years later, Davisson and Germer experimentally observed that electrons underwent diffraction just as light did. These were landmark moments in the history of physics – de Broglie received the Nobel Prize in physics (the second time it was awarded for a PhD thesis) and, later, so did Davisson and Germer. While the Davisson–Germer experiment was the first to establish the dual nature of matter, the double slit interference experiment is far easier to conceptually grasp and visualise, which is why we will use it to embark on our study of quantum phenomena.

The setup of an electron interferometer used in the double slit experiment is conceptually quite similar to that of a light interferometer (Figure 1.1). A parallel beam of electrons is incident on a screen with two slits. The electrons that pass through the slits impinge upon the optical screen, where their incidence is captured by a visible spot. First, let us think of these electrons as if they were the kind of particles we observe in our daily lives (classical particles) and see how they should behave. All the electrons in the initial beam have the same speed and direction of motion and they are heading towards the screen with two slits. All the electrons that hit this screen are blocked, except for the ones passing right through the slits. These electrons that passed through the slits should have no reason to change either their speed or the direction of their motion. They do not "know" that there

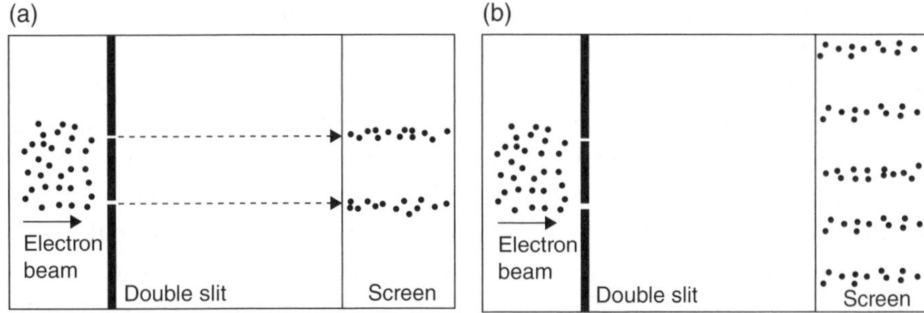

Figure 1.1 (a) Classically predicted electron pattern. It can be seen that interference fringes are experimentally observed, as opposed to the classically predicted pattern. This establishes the wave-like behaviour of electrons, (b) experimentally observed electron pattern.

was a screen in the first place – they pass through unaffected. They subsequently keep heading straight and hit the final screen as illustrated in Figure 1.1(a). Two narrow bands are formed on the screen, corresponding to the two thin beams of electrons that passed unaffected through the two small slits.

Now let us take a look at what was actually observed in the experiment. There was an interference pattern on the screen, as shown in Figure 1.1(b), a pattern uncannily similar to what is observed when we perform the same experiment with light instead of electrons. At this juncture, you might hypothesise this behaviour to result from some sort of statistical phenomenon due to the large number of electrons. However, the experiment is far from finished, and further strangeness lies ahead.

Let us now adjust the electron source so that instead of a beam of electrons it sends a single electron at a time. This time, we find something even more extraordinary – after a lot of electrons have hit the screen, the same interference pattern builds up as in the case of a beam of many electrons. There is no way this electron "knows" that it has been preceded by, or it will be followed by, another electron. What, then, could be happening? The answer is even more puzzling than the question, and will take you quite a while to come to terms with – *each and every electron is undergoing interference with itself.* This is what leads to the final conclusion that not just aggregates of particles but each and every particle exhibits a wave nature. To make this point clear, let us modify the experiment such that we are able to find out through which slit each electron passes. Independent of *how* we find out which slit each electron passes through, we get *exactly* the same result, that is the interference pattern vanishes and we get the pattern shown in Figure 1.1(a), as predicted by classical mechanics. Think about this very carefully, because this point merits serious investigation. For a wave to show interference, there need to be two sources – the two slits in this case. Thus, for a single electron to show interference, *it must be passing through both slits*. However, this is not possible! At the very least, we cannot imagine such a situation. It is only reasonable to assume that the electron either goes through one slit or the other, but the moment we impose such a restriction on the electron, we are thinking of it as a classical particle. By just *knowing* which slit the electron is going through, and thereby imposing the condition that it will pass through either one slit or the other, we are restricting it to behave like a classical particle. While the mathematical foundations will be laid later in this chapter, for now the reader should try and grasp the underlying concept – the quantum electron passes through *both* the slits; it is a *superposition* of these two states (corresponding to passing through the upper or lower slit). You may think that the electron *actually* passes through either of the two slits and due to limitations of our experimental techniques, we do not know which slit it passes through. This is not the case – the electron is indeed passing through both the slits. This counterintuitive phenomenon is at the very root of quantum mechanics and it will take some time for us to be familiar with this kind of approach. You *cannot* ask of the *quantum* electron (or any general quantum particle), "Which slit does it pass through?" The question in itself is wrong. It passes through both. It should be noted that this wave nature of a particle becomes appreciable only at very small sizes, such as a few nanometres.

1.1.2 Basic concepts of quantum mechanics

1.1.2.1 Wavefunctions

The behaviour of classical particles can be fully explained by describing how their position changes with time. This information would be sufficient to give us the trajectory, the velocity, the momentum and the acceleration of the particle. However, what of the quantum particle? Surely, the electron that passed through both slits of the double slit experiment cannot be assigned a precise location. This leads us to the realisation that we need some new method to describe the quantum particle. The rest of this chapter is devoted to formulating a mathematical picture that is able to capture the unusual behaviour of quantum particles.

The search for this new method of description was helped by the knowledge that the quantum behaviour of particles closely resembled the behaviour displayed by waves. Waves of many kinds – electromagnetic waves, sound waves, etc. – had been extensively studied, and all these waves were described by wave equations. These equations described the behaviour of a wave at every point in space, and at all times. For example, in the case of sound waves, the wave equation described the displacement ($\Delta\vec{r}$) of each particle as a function of time:

$$\Delta\vec{r} = \vec{\psi}\,(x,y,z,t) \tag{1.1}$$

Similarly, for an electromagnetic wave, the wave equations described the electric ($\vec{E}$) or magnetic ($\vec{B}$) field at each and every point as a function of time:

$$\vec{E} = \overrightarrow{\psi_E}\,(x,y,z,t) \tag{1.2}$$

$$\vec{B} = \overrightarrow{\psi_B}\,(x,y,z,t) \tag{1.3}$$

Taking the cue from these equations, physicists assigned a similar wave equation to the quantum particle. This equation was called the wavefunction of the quantum particle and was usually denoted by the Greek symbol ψ (psi). Just like in the case of classical waves, this wavefunction contained all the information about the particle – its current state and the variation of its behaviour with time. It is important to note that while the previously described wave equations (1.1) to (1.3) were real functions, the wavefunction of a quantum particle is a complex function.

1.1.2.2 Born interpretation

While the wavefunction-based formulation of quantum mechanics was proposed by Erwin Schrodinger quite early, he was at a loss to ascribe any physical meaning to it. The theory he built up described what sort of mathematical operations one needed to perform on the wavefunction to get information relating to its various properties, such as its position, its momentum, its energy, etc. However, what this wavefunction *itself* meant was a mystery, especially because it was a complex function. Many interpretations were proposed as to the meaning of this wavefunction, but the one that is most widely accepted was proposed by Max Born. It is known as the "Born interpretation of quantum mechanics" and is one of the fundamental principles of quantum mechanics. According to this interpretation, the wavefunction ψ is the "probability amplitude" of the quantum particle, the square of whose magnitude gives us the probability density ρ of finding that particle at any point:

$$\rho = |\psi|^2 = \psi^*\psi \tag{1.4}$$

where $\psi*$ is the complex conjugate of ψ. Using this interpretation, the probability P of finding the particle in a volume V at any time t would be

$$P(V,t) = \iiint_V \rho \; dx\; dy\; dz \tag{1.5}$$

Since the probability of finding the particle in the entire space should always be unity, we can say that

$$\iiint_{-\infty}^{\infty} \psi^* \psi \, dx \, dy \, dz = 1 \tag{1.6}$$

A wavefunction that displays this property is called a "normalised" wavefunction.

It is very important to realise that the probabilistic behaviour that follows from the Born interpretation is different from the probabilistic behaviour encountered in statistical mechanics. For example, consider an ensemble of particles in a chamber each occupying a particular position. This allows us to calculate the probability of finding a particle at any position. If there were only a single classical particle in this chamber, we could always precisely identify its position. However, in quantum mechanics, every single particle is "spread out" in space, and its position is uncertain. Even if there is only a single electron, we cannot say, "The electron is at this particular point". We can only talk about the probability of finding the electron at any given point once we measure its position. Thus, the Born interpretation provided a physical meaning to the wavefunction that was compatible with the fundamentally probabilistic behaviour of a quantum particle, and gave a mathematical approach to calculate the probability of finding a quantum particle at any region in space.

1.1.2.3 Measurement

Measurement is a fundamental process in our lives, yet it is so much a part of our instincts that we barely pay any attention to it. However, if you think carefully, most of the information you get is by the process of measurement. When you look at a tree, your eyes measure the frequency and amplitude of the incoming electromagnetic waves, giving you information regarding the colour and brightness of the tree. Subsequently, your eyes measure the angular difference between the signals received by the two eyes, and calculations by your brain tell you how far away this tree is. You may hear a bird chirping on this tree – once again, due to your ears measuring the frequency and location of the pressure waves (sound) impinging upon them. Similarly, any information we get about a quantum particle is by the process of measurement – measuring the position, energy, momentum, etc.

However, there is a very fundamental difference between measurement in classical physics and quantum physics. While the state of a classical particle is independent of measurements performed on it, in quantum mechanics, the state of the quantum particle is intricately linked to measurements performed on it. We shall go back to the double slit experiment to illustrate this point. When we measured which slit the electron passed through, that is when we measured its position, it stopped showing interference. The electron, before measurement, exhibited interference. After we carried out the measurement, it no longer showed interference. This shows that *measurement changed the state of the electron.* In general, measurement changes the state of a quantum particle, and its final state (after measurement) depends both on its initial state and the kind of measurement being performed. Do not be worried if the picture is not completely clear yet – to fully understand the process of measurement, we will have to know about operators and eigenvalues, which we will do in the next two sections.

1.1.2.4 Operators

The Born interpretation told us that we can obtain the probability of finding a quantum particle at any given point if we know its wavefunction. However, the wavefunction contains far more information than this. If you remember, the wavefunction was supposed to contain *all* the information about the quantum particle. How, then, do we extract this information from the wavefunction?

Since the wavefunction is a mathematical function, it is clear that we will be performing certain mathematical operations on it to get the information we desire. This mathematical operation must be different, depending on the specific kind of information – energy, momentum, position, etc. – we need to obtain. This, indeed, is the case.

Corresponding to every physically observable parameter (also called observables) of a quantum particle, such as position, momentum, energy, we have mathematical operators. The operators for certain common observables are listed below in Table 1.1, where ι (iota) is the square root of negative unity and $\hbar$ (h-cross or h-bar) is the reduced Planck's constant.

To understand the use of these operators, let us imagine an experiment where we have a large number of quantum particles with the same wavefunction ψ. We wish to measure a particular observable, the mathematical operator corresponding to which is O. The outcome of each measurement is o. As the behaviour of

Table 1.1 Quantum mechanical operators corresponding to physical observables.

Observable	Operator
Position (x)	x
Momentum ($\vec{p}$)	$-\iota\hbar\vec{\nabla}$
Energy (E)	$\iota\hbar\frac{\partial}{\partial t}$

quantum particles is probabilistic, measuring O for every particle will give a different outcome o. Looking back to our example of the double slit experiment, all the incoming electrons were exactly similar. However, when we start measuring which slit they pass through, sometimes we find that an electron passes through the upper slit and at other times through the lower slit. We can, therefore, only discuss the expectation value $\langle o \rangle$ after taking an average of all the measurements. This expectation value of the observable o is given as

$$\langle o \rangle = \frac{\int\int\int_{-\infty}^{\infty} \psi^* O \psi \, dx \, dy \, dz}{\int\int\int_{-\infty}^{\infty} \psi^* \psi \, dx \, dy \, dz} \tag{1.7}$$

The above equation tells us about the expectation value when we perform a large number of measurements, all on particles with the same wavefunction ψ. However, if we have only one particle, it would be useful to know the probability of obtaining a particular result. For us to know this, we must find the eigenfunctions of the operator in question.

1.1.2.5 Eigenfunctions

Let us recollect from the section on measurement (Section 1.1.2.3) that the state of a quantum particle changes upon measurement, and the final state is dependent on both the initial state and the kind of measurement being performed. However, there are certain very special states corresponding to every observable that do not change when it is measured. These special states are the eigenfunctions of that observable. If, for an operator O, the wavefunction ψ_o behaves as

$$O\psi_o = \lambda \psi_o \tag{1.8}$$

where λ is a constant, then ψ_o is an eigenfunction (also referred to as an eigenstate or an eigenvector) of the operator O and λ is the corresponding eigenvalue. Suppose that we measure the observable corresponding to the operator O on a particle having the wavefunction ψ_o. We will find the value of this observable to be λ. This can be proven by substituting the value of $O\psi_o$ from Equation (1.8) into Equation (1.7) that gave us the expectation value corresponding to any operator. Moreover, the wavefunction ψ_o will remain unchanged. Therefore, for a particle whose wavefunction is an eigenfunction of an observable, we can, with absolute certainty, state the result of measurement. To understand this, let us consider the energy operator (Table 1.1) as an example. Let us assume $\psi_{E_i} (i = 0, 1, 2, \ldots)$ to be the eigenfunctions of the energy operator, having eigenvalues E_i:

$$\imath\hbar \frac{\partial \psi_{E_i}}{\partial t} = E_i \psi_{E_i} \tag{1.9}$$

If we take a particle with wavefunction ψ_{E_i}, we *know* that its energy is E_i. There is no probability involved in this.

Furthermore, even if we have a wavefunction that is not an eigenfunction of the operator in question, the result of every measurement can *only* be one of the eigenvalues of the operator. Let us again take the energy operator as an example. Suppose that we measure the energy of a particle having a wavefunction Ψ that is *not* one of the eigenfunctions ψ_{E_i} of the operator. The result will *always* be one of the eigenvalues E_i. Note that every measurement will result in a *different* energy being observed each time. You will now say that once you have measured the energy, and it is found to be a particular E_i, you *know* the energy of the particle to be E_i. However, it was stated earlier in this section that we can only know (with absolute certainty) the energy of the eigenfunctions of an observable. Yet Ψ is not an eigenfunction of the energy operator. What happens is that, after measurement, the wavefunction Ψ "collapses" into the wavefunction ψ_{E_i} corresponding to the observed energy E_i. Remember that measurement changes the state of a quantum particle. Now we can say that *measuring an observable leads us to observe one of the eigenvalues of that observable, and the state of the quantum particle being measured changes to the corresponding eigenfunction.*

We now face the problem of finding the probability of this "collapse" into a particular eigenfunction. Linear algebra provides us with a very handy solution to this problem. Any general wavefunction can be written in terms of the eigenfunctions of an operator. Let us clarify this point. Every operator has a set of eigenfunctions. If we use a linear combination of all these eigenfunctions, we get a set of states that includes every possible state that the quantum particle can have. That is, every wavefunction can be decomposed into a linear superposition of the eigenfunctions of any given operator. Let us once again go back to the energy operator and its eigenfunctions that we discussed in Equation (1.9). *Any* general wavefunction Ψ can be written in terms of the eigenfunctions ψ_{E_i} of the energy operator as

$$\Psi = \Sigma a_i \psi_{E_i} \tag{1.10}$$

where a_i are coefficients corresponding to every wavefunction ψ_{E_i} and are complex numbers.

Unlike in the case of the eigenfunctions ψ_{E_i}, we cannot discuss the energy of this general particle as having wavefunction Ψ. It is a linear superposition of states ψ_{E_i} of different energies E_i. When the energy of this particle is measured, one obtains any one of the energies E_i, and the particle is found to be in the state ψ_{E_i} after the measurement. However, we cannot say that the particle had energy E_i because the measurement may very well have led to the observation of a different energy E_j. Now the state of the particle changes after measurement from

Ψ to ψ_{E_i}. The probability $P(\psi_{E_i})$ that the wavefunction Ψ collapses into a particular eigenfunction ψ_{E_i} can be written as

$$P(\psi_{E_i}) = |a_i|^2 \tag{1.11}$$

where a_i is the coefficient corresponding the eigenfunction ψ_{E_i} in the linear superposition shown in Equation (1.10).

1.1.3 Schrodinger's equation

1.1.3.1 Formulation of the equation

We have now understood what wavefunctions mean and how they behave when measured, yet we do not know how to find them for a particular physical situation, like an electron in a hydrogen atom. There must be some equations that have to be solved to give us these wavefunctions. Just as Maxwell's equations (when solved under an appropriate set of boundary conditions) give the equations for electromagnetic waves, an equation is needed that can be used to find the wavefunction of a quantum particle. This equation is called Schrodinger's equation. It is a quantum formulation of the statement that

$$Total\ \ energy = kinetic\ \ energy + potential\ \ energy \tag{1.12}$$

By using the operators listed in Section 1.1.2.4 (Table 1.1), we can write:

$$Kinetic\ \ energy = \frac{p^2}{2m} = -\frac{\hbar^2}{2m}\nabla^2 \tag{1.13a}$$

$$Potential\ \ energy = V(\vec{r},t) \tag{1.13b}$$

$$Total\ \ energy = \imath\hbar\frac{\partial}{\partial t} \tag{1.13c}$$

which, when substituted into Equation (1.12), gives Schrodinger's equation:

$$-\frac{\hbar^2}{2m}\nabla^2 + V(\vec{r},t) = \imath\hbar\frac{\partial}{\partial t} \tag{1.14}$$

The above equation is in the form of operators of the individual energies. It needs to be operated upon the wavefunction ψ, giving the final form of the equation as

$$-\frac{\hbar^2}{2m}\nabla^2\psi + V\psi = \imath\hbar\frac{\partial\psi}{\partial t} \tag{1.15}$$

The above equation is a partial differential equation, which has to be solved for ψ. The form of the potential $V(\vec{r},t)$ and the boundary conditions will be different for different physical problems (e.g. a particle confined in a one-dimensional well, the tunnelling problem, the hydrogen atom, etc.), thus leading to different wavefunctions.

In Equation (1.15), we have introduced the time-dependent form of Schrodinger's equation. However, in most cases, we would be solving for energy eigenfunctions, which are stationary states and do not change with time. Therefore, the right-hand side of Equation (1.15) changes to $E\psi$, giving

$$-\frac{\hbar^2}{2m}\nabla^2\psi + V\psi = E\psi \tag{1.16}$$

This is the time-independent form of Schrodinger's equation, and it will be the building block for solving most elementary and slightly complex quantum problems. Additionally, most of the problems that are dealt with in this text are one dimensional. In this case, the time-independent Schrodinger equation (1.16) further simplifies to

$$-\frac{\hbar^2}{2m}\frac{\partial^2\psi}{\partial x^2} + V\psi = E\psi \tag{1.17}$$

Using Schrodinger's equation, you can get some very important mathematical conditions that every wavefunction must obey [1]. These are:

1. The wavefunction must be continuous at each point in space.
2. The first derivative of the wavefunction must be continuous at each point in space, unless the potential V at the boundary in question is infinite.

1.1.3.2 Probability current

Our final aim is to model tunnel field-effect transistors (TFETs) for predicting their electric currents. We must keep this perspective in mind when we look at the quantum mechanical techniques that we are discussing. We know that electric current is a measure of the rate of flow of charge. These charges are either electrons or holes and their behaviour is best predicted by quantum mechanics. Therefore, we need to find a link between the electric current and the quantum mechanical behaviour of charge carriers.

Let us consider a single electron present inside a conducting wire. If we were to study the behaviour of this electron purely in terms of classical physics, we could find its position and velocity, and use this information to calculate the current in

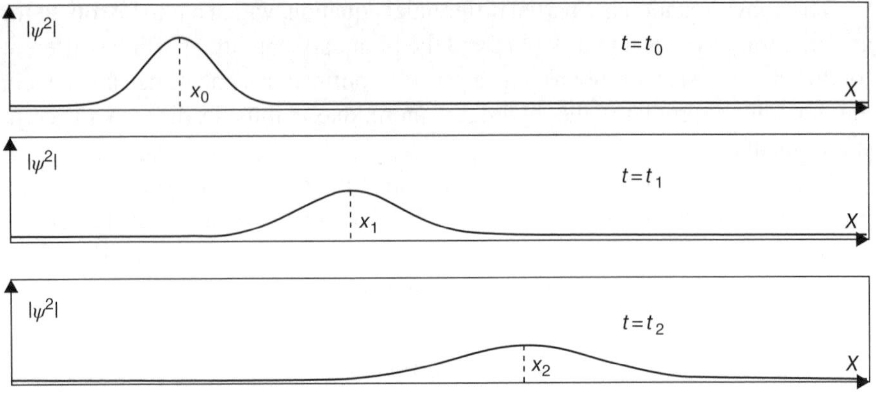

Figure 1.2 Probability density of a quantum particle moving from left to right, plotted at different points of time ($t_2 > t_1 > t_0$).

the wire. However, for an electron obeying the laws of quantum mechanics, we cannot specify its position – we can only find the probability of its presence at any point. From this point of view, what happens when an electron moves in a particular direction? The probability density $|\psi^2|$of a quantum particle moving from left to right is plotted at various points of time in Figure 1.2. In terms of quantum mechanics, we might say that the probability of finding this particle is changing. In technical terms, there is now a probability current, which is the rate of flow of the probability of the quantum particle's presence. To calculate the electric current resulting from the flow of quantum particles, we need to find this probability current.

We know that the differential form of the continuity equation for charge is

$$\vec{\nabla} \cdot \vec{J} = -\frac{\partial \rho}{\partial t} \tag{1.18}$$

where $\vec{J}$ is the current density and ρ is the charge density. In the case of a quantum particle, the same continuity equation holds, but with $\vec{J}$ being the probability current and $\rho = \psi^* \psi$ being the probability density. We can use the left-hand side of the time-independent Schrodinger Equation (1.16) to give us the values of ψ and $\psi*$. When these values of ψ and $\psi*$ are substituted in Equation (1.18), we can write the probability current $\vec{J}$ as [1]

$$\vec{J} = \frac{\imath\hbar}{2m}\left(\psi \vec{\nabla} \psi^* - \psi^* \vec{\nabla} \psi\right) \tag{1.19}$$

1.2 Basic quantum physics problems

In this section, we will familiarise ourselves with certain basic problems of quantum mechanics. This would enable us to appreciate the concepts introduced that act as a bridge between the theoretical framework and the practical problem of tunnelling.

1.2.1 Free particle

The simplest quantum particle is a free particle – a particle that is completely unconstrained by any external potentials. This free particle is a basic building block for studying more complex quantum mechanical problems, just as we use the sine wave as a basic building block for representing complicated waves (by using Fourier series).

1.2.1.1 Wavefunction

As a free particle is unconstrained by any external potential, the term $V(\vec{r})$ in the time-independent Schrodinger equation (1.16) is zero at all points in space. Therefore, for a free particle, we can write

$$-\frac{\hbar^2}{2m}\frac{\partial^2\psi}{\partial x^2}=E\psi \tag{1.20}$$

The solutions of Equation (1.20) are

$$\psi_{\pm}=Ae^{\pm\imath kx} \tag{1.21}$$

$$k=\sqrt{\frac{2mE}{\hbar^2}} \tag{1.22}$$

where k is referred to as the wave vector of the free particle and A is a complex constant. The general solution would be a linear superposition of ψ_+ and ψ_-:

$$\Psi=A'e^{\imath kx}+B'e^{-\imath kx} \tag{1.23}$$

where A' and B' are complex constants.

We find that there are two eigenvectors ψ_+ and ψ_- corresponding to a particular energy E – one with a plus sign and another with a minus sign in the exponential. Let us see what happens when we apply the momentum operator p_x (Table 1.1) to these wavefunctions:

$$p_x\psi_{\pm}=-\imath\hbar\frac{\partial\psi_{\pm}}{\partial x}=\pm\hbar k\psi_{\pm} \tag{1.24}$$

We find that the wavefunctions $\psi_{\pm}$ are eigenfunctions of the momentum operator, with eigenvalues $\pm \hbar k$. This means that the particle corresponding to ψ_{+} has momentum in the positive x-direction, while ψ_{-} has momentum in the negative x-direction. Thus, ψ_{+} is a free particle moving forward with a momentum $\hbar k$ and ψ_{-} is a free particle moving backward with a momentum $\hbar k$.

Another important observation here is that the probability ρ (Equation (1.4)) of finding the particle at any location is uniform – the particle is fully dispersed in space. Therefore, we have

$$\rho = \psi^* \psi = |A|^2 \tag{1.25}$$

The fact that the particle is fully dispersed in space follows from Heisenberg's uncertainty principle. The momentum of either of the two wavefunctions $\psi_{\pm}$ is known with complete certainty. Therefore, the uncertainty in the position of the particle is infinite.

1.2.1.2 Probability current

Using Equation (1.19), we can write the probability current for a free particle propagating along the positive x-axis as

$$J = \frac{\hbar k}{m} |A|^2 \tag{1.26}$$

Since $|A|^2$ is the probability density (Equation (1.25)) and $\hbar k$ is the momentum (Equation (1.24)) of the particle, the above equation can be rearranged to give

$$J = \frac{p}{m} \rho = v\rho \tag{1.27}$$

where $v = p/m$ is the velocity of the particle. We can now write the electric current density j as

$$j = qJ = qv\rho \tag{1.28}$$

In this section, we have discussed the behaviour of a completely unconstrained "free" particle. Let us now investigate the behaviour of a quantum particle that is bound to remain between two points – a particle in a box.

1.2.2 Particle in a one-dimensional box

The next problem we will deal with is a particle confined to move along an axis between two points. It cannot go beyond those two specified points. It is, therefore, referred to as a particle in a one-dimensional box. It is confined to remain within the

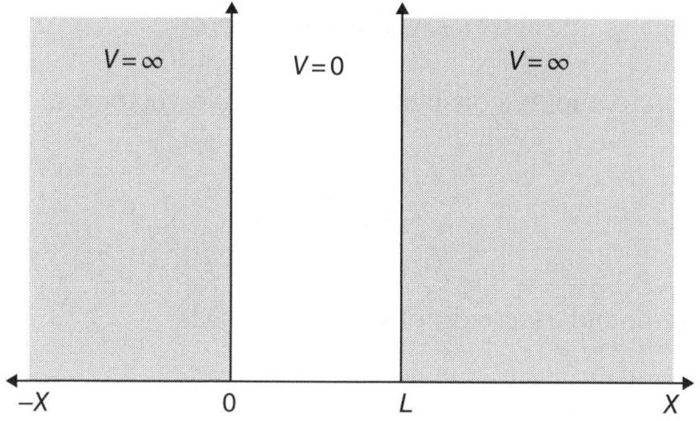

Figure 1.3 Particle in a one-dimensional box.

"box" (also referred to as an infinite potential well) with edges at $x=0$ and $x=L$, as shown in Figure 1.3. The particle is completely free to move within these limits, implying that there is no external potential (i.e. $V=0$) in the region $x\epsilon(0,L)$. Outside these limits, there is an infinite potential barrier (i.e. $V=\infty$). Physically, this form of external potential ensures that the particle stays within the "box". Since Schrodinger's equation assumes different forms in different regions, its solution – the wavefunction – will also have different forms in different regions. We will, therefore, find the wavefunction separately for each region and then join these separate solutions to give the final solution. It is important to be familiar with this procedure since it will be used in this book, not just for solving Schrodinger's equation but as a very important tool in the modelling of TFETs.

Now we know that the particle cannot be present in the regions $x\le 0$ and $x\ge L$. Thus, in these regions, the probability of finding the electron is zero:

$$\rho=\psi^*\psi=0 \tag{1.29}$$

This is only possible if $\psi=0$ in these regions.

Let us now find the solution of Schrodinger's equation within the box, where the form of Schrodinger's equation is the same as that of the free particle (Equation (1.20)). However, we need to keep in mind the restrictions applied to a wavefunction, stated at the end of Section 1.1.3.1. The first condition is that the wavefunction must be continuous. Since the wavefunction is zero in the regions $x\le 0$ and $x\ge L$, we get the following boundary conditions at the edges of our box:

$$\psi(0)=0 \tag{1.30a}$$

$$\psi(L)=0 \tag{1.30b}$$

The general solution of Equation (1.20) shown in Equation (1.23) can be written in the form of trigonometric functions as

$$\psi(x) = A\ \sin kx + B\ \cos kx \tag{1.31}$$

where A and B are complex constants and the wave vector k is

$$k = \sqrt{\frac{2mE}{\hbar^2}} \tag{1.32}$$

Applying the boundary conditions (1.30), we get

$$B = 0 \tag{1.33}$$

$$k_n L = n\pi \tag{1.34}$$

The subscript n has been added to k because Equation (1.34) shows that k is quantised and takes only certain fixed values. Using Equation (1.34) in (1.32) gives the possible values of energy E_n that the particle can have:

$$E_n = \frac{n^2 h^2}{8mL^2} \tag{1.35}$$

and the wavefunction ψ_n corresponding the energy E_n is

$$\psi_n = A\ \sin\frac{n\pi x}{L} \tag{1.36}$$

The constant A is given by the normalisation condition (1.6):

$$A = \frac{1}{\sqrt{2L}} \tag{1.37}$$

The above wavefunctions (1.36) correspond to energy eigenstates, as we have solved the time-independent Schrodinger equation. A general wavefunction for a particle in a one-dimensional box would, therefore, be a superposition of these energy eigenstates:

$$\Psi = \Sigma a_n \psi_n \tag{1.38}$$

where a_n is the coefficient corresponding to the wavefunction ψ_n and is a complex number.

If we plot the wavefunctions of energy eigenstates for a particle in a one-dimensional box (Figure 1.4), we observe the same pattern as standing waves in a string.

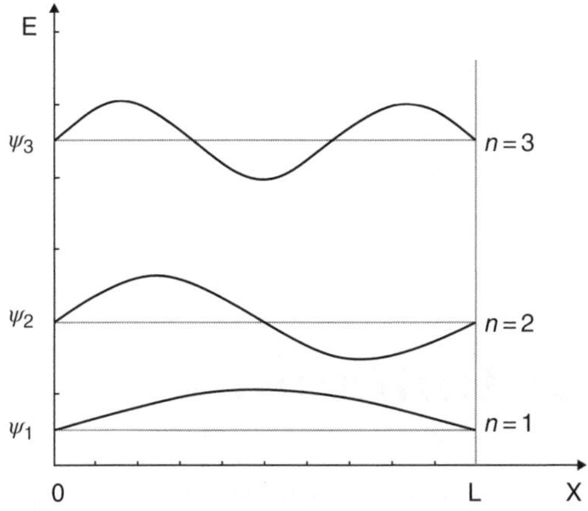

Figure 1.4 Energy eigenfunctions of a particle in a one-dimensional box.

Going back to classical physics, remember that a standing wave can be written as a superposition of two travelling waves. Therefore, we can represent the energy eigenstate of a particle in a one-dimensional box as a superposition of two free particle wavefunctions going in opposite directions:

$$\psi_n = \frac{1}{\sqrt{2L}} \sin k_n x = \frac{\imath}{\sqrt{2L}}\left(e^{-\imath k_n x} - e^{\imath k_n x}\right) \tag{1.39}$$

The first and second exponential terms in Equation (1.39) represent a free particle travelling in the negative and positive x-directions with momentum $\hbar k_n$, respectively. Therefore, if we take the expectation value (1.7) of their superposition, these momenta cancel each other and the expectation value of the momentum for ψ_n is zero.

Reference

[1] D. J. Griffiths, *Introduction to Quantum Mechanics*, 2nd edn, Pearson Education, 2005.

2

Basics of tunnelling

2.1 Understanding tunnelling

In the previous chapter, we developed the necessary theoretical framework to deal with the problem of quantum mechanical tunnelling. Tunnelling plays an important role in many physical phenomena and devices – nuclear fusion, alpha decay, the scanning tunnelling microscope, the tunnel diode, tunnelling FETs, etc. The focus of this book is the modelling of *tunnelling* field-effect transistors. As is evident from the name, tunnelling is at the very basis of what makes TFETs work.

Let us first develop a qualitative idea of tunnelling, after which we will examine its quantitative details.

2.1.1 Qualitative description

Tunnelling is a quantum phenomenon, in which a particle is able to cross a potential barrier even though it does not have the energy to overcome this barrier. Such behaviour is not observed in the case of classical particles. Therefore, any classical analogy used to explain the phenomenon of quantum mechanical tunnelling would necessarily be inaccurate. Instead of taking such a classical analogy, it would be more fruitful to picture the quantum particle as a wave and form a link between this quantum mechanical behaviour and the behaviour displayed by waves.

Let us consider the problem of an electromagnetic wave incident on a conductor. The equations that describe this phenomenon have the same form as

Tunnel Field-Effect Transistors (TFET): Modelling and Simulation, First Edition. Jagadesh Kumar Mamidala, Rajat Vishnoi and Pratyush Pandey.

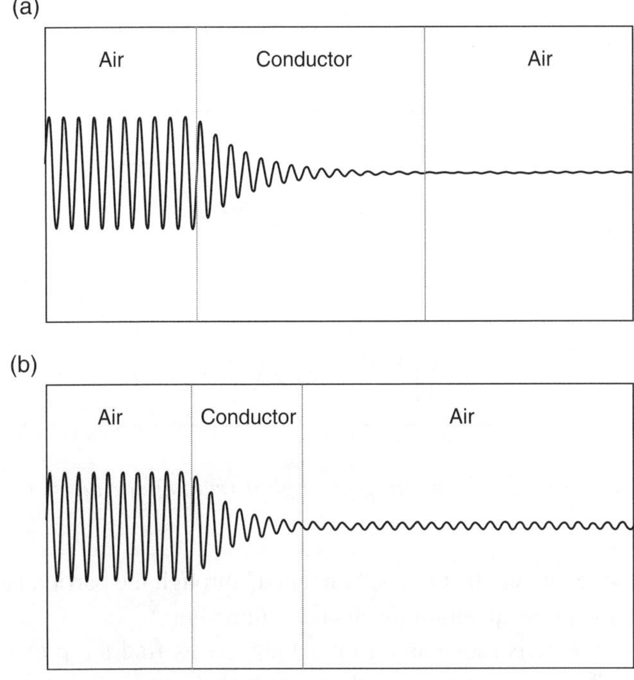

Figure 2.1 (a) Emerging electromagnetic wave has negligible amplitude. (b) Emerging electromagnetic wave's amplitude is attenuated, but not to negligible levels.

the equations that describe quantum mechanical tunnelling. When an electromagnetic wave is incident on a conductor, the free electrons of the conductor move. This causes an energy transfer – the wave loses energy and the electrons gain energy. Consequently, the amplitude of the wave decays exponentially within the material and is soon negligible, as shown in Figure 2.1(a). However, if the conductor is very thin, the amplitude of the wave does not decay to negligible levels before reaching the other edge of the conductor. In this case, we find electromagnetic waves of a very small amplitude on the other side of the conductor, as shown in Figure 2.1(b). In the previous chapter, we represented quantum particles by wavefunctions whose amplitudes correspond to the probability of the particle's presence. If we substitute the electromagnetic waves by such quantum particles and the conductor by a potential barrier, we would get a strikingly similar situation. A quantum mechanical particle is incident upon a potential barrier that it does not have enough energy to cross, yet there is a non-zero probability of the particle being found on the other side of this barrier. If we observe the particle

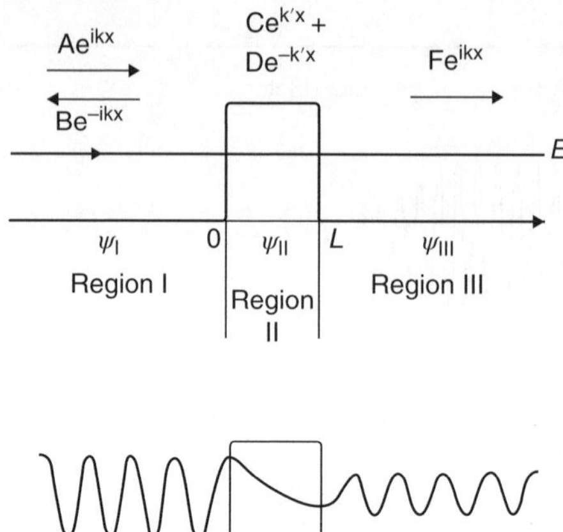

Figure 2.2 Tunnelling through a rectangular barrier.

on the other side, we say that it has "tunnelled" through the barrier, and this phenomenon is known as quantum mechanical tunnelling.

With this qualitative idea about tunnelling, let us find the probability of an electron tunnelling through a potential barrier. In the case of an electron, the potential barrier is the external potential $V(x)$ term of Schrodinger's equation. This function can take various forms. Let us first deal with the relatively simple problem of tunnelling through a rectangular potential barrier, after which we shall develop a method to obtain the probability of tunnelling through any potential barrier.

2.1.2 Rectangular barrier

The simplest tunnelling problem is that of a rectangular potential barrier, for which the external potential term $V(x)$ in Schrodinger's equation is

$$V(x)=\begin{cases}0, x\leq 0\\ V, x\in(0,L)\\ 0, x\geq L\end{cases} \tag{2.1}$$

As we did in the case of a particle in a one-dimensional box (Section 1.2.2), we divide the space into three regions as shown in Figure 2.2. Region I ($x\leq 0$) is where the incoming particle impinges upon the barrier. Region II ($x\in(0,L)$) is the barrier itself. Region III ($x\geq L$) is where the particle emerges after tunnelling through the barrier.

Within Region I, the potential $V=0$; hence Schrodinger's equation reduces to that of a free particle (1.20). The incoming particle moving towards the right in

Region I, therefore, has a wavefunction $\psi_{I,i} = Ae^{ikx}$. Notice that we chose the ψ_+ solution of a free particle (1.21) to represent the incoming particle since it is moving along the positive x-axis and therefore has a momentum in the positive x-direction.

At this point, we need to start looking at the particle as if it were a wave. When a wave hits the edge of a barrier, a part of it is reflected and a part of it is transmitted. Therefore, this incoming wave-particle is partially reflected, resulting in another free particle moving towards the left in Region I having a wavefunction $\psi_{I,r} = Be^{-ikx}$. Once again, notice that we chose the ψ_- solution from Equation (1.21) because the reflected particle is now moving to the left and thus has momentum in the negative x-direction.

Superposing the wavefunctions of the incoming and the reflected particle, the resultant wavefunction in Region I is

$$\psi_I = \psi_{I,i} + \psi_{I,r} = Ae^{ikx} + Be^{-ikx} \tag{2.2}$$

This wavefunction ψ_I is, of course, the general solution (Equation (1.23)) to Schrodinger's equation for a free particle. We chose to arrive at the same solution in a circuitous way so that we could assign a physical meaning to the eigenfunctions ψ_+ and ψ_-, which represent the incoming and the reflected particles, respectively.

Let us now write Schrodinger's equation in Region II:

$$\frac{\hbar^2}{2m}\frac{\partial^2 \psi_{II}}{\partial x^2} = (V - E)\psi_{II} \tag{2.3}$$

where ψ_{II} is the wavefunction of the particle in Region II. If the energy E of the particle is more than the barrier height V (i.e. $V < E$) we know that the particle can cross over the barrier. However, suppose that the energy E of the particle is less than what is needed to cross over the barrier height V ($V > E$), and the particle still moves from Region I to Region III. This phenomenon is called quantum mechanical tunnelling and, in this case, the solution for Equation (2.3) is

$$\psi_{II} = Ce^{\pm k'x} \tag{2.4}$$

$$k' = \sqrt{\frac{2m(V - E)}{\hbar^2}} \tag{2.5}$$

The general solution in Region II, therefore, is

$$\psi_{II} = Ce^{k'x} + De^{-k'x} \tag{2.6}$$

where C and D are complex constants.

In Region III, the potential $V=0$ and the transmitted particle is moving towards the right and has a positive momentum. Therefore, we can use the free particle solution (1.23) of Schrodinger's equation, without the negative momentum $e^{-\imath kx}$ term, to write the wavefunction ψ_{III} in Region III:

$$\psi_{III} = Fe^{\imath kx} \tag{2.7}$$

where F is a complex constant.

Since the wavefunctions in Regions I, II and III are known, the combined wavefunction Ψ can be written as

$$\Psi = \begin{cases} \psi_{I,i} + \psi_{I,r}, & x \leq 0 \\ \psi_{II}, & x \in (0, L) \\ \psi_{III}, & x \geq L \end{cases} \tag{2.8}$$

We only need to solve for the constants A, B, C, D, and F to get the final wavefunction Ψ. Once again, we use the physical conditions of continuity and differentiability of the wavefunction at all points to obtain these constants. Within each region, the wavefunctions are continuous and differentiable. We need to impose these boundary conditions only at the points where the wavefunction Ψ changes its form, that is at $x=0$ and $x=L$.

Imposing continuity and differentiability, respectively, of the wavefunction Ψ at $x=0$, we get

$$A + B = C + D \tag{2.9}$$

$$\imath k(A - B) = k'(C - D) \tag{2.10}$$

Similarly, at $x=L$, we get

$$Ce^{k'L} + De^{k'L} = Fe^{\imath kL} \tag{2.11}$$

$$k'\left(Ce^{k'L} - De^{-k'L}\right) = \imath kFe^{\imath kL} \tag{2.12}$$

Notice that we have five variables but only four linear equations. To find the tunnelling probability, we do not need to find the value of all the five variables but only the ratio of two variables. The probability density of the transmitted particle (in Region III) is $\rho_{III} = |F|^2$ and the probability density of the incoming particle (in Region I) is $\rho_{I,i} = |A|^2$. The tunnelling (or transmission) probability is, therefore, given by

$$T = \frac{|F|^2}{|A|^2} \tag{2.13}$$

Using Equations (2.9) to (2.12) to solve for the tunnelling probability given in (2.13), we get

$$T = \frac{1}{1 + \dfrac{V^2 \sinh^2 k'L}{4E(V-E)}} \tag{2.14}$$

This formula (2.14) is useful because some potential barriers may be approximated as rectangular barriers. However, many practical tunnelling problems cannot be approximated as tunnelling through a rectangular potential barrier. Therefore, for practical tunnelling problems, we need to develop a method that can give us the tunnelling probability for any type of potential barrier. Such a method is described in the next section.

2.2 WKB approximation

For a rectangular potential barrier (Section 2.1.2), the external potential $V(x)$ in Schrodinger's equation was constant, thereby simplifying the process of obtaining the solution. However, finding the tunnelling probability through a general potential barrier involves solving Schrodinger's equation where the external potential $V(x)$ can have any form. This is mathematically challenging and therefore we need to apply approximations to simplify the problem. One such approximation is the WKB approximation, which is widely used to find the tunnelling probability [1]. It was developed by Wentzel, Kramers and Brillouin, and is a mathematical approach to find approximate solutions to certain types of differential equations. We will now discuss this approach by applying it to the problem of quantum mechanical tunnelling.

Let us begin by writing Schrodinger's equation for a general potential barrier $V(x)$:

$$-\frac{\hbar^2}{2m}\frac{\partial^2 \psi}{\partial x^2} + V(x)\psi = E\psi \tag{2.15}$$

Equation (2.15) is a partial differential equation and we will use the WKB approximation to solve it. There is little physics involved in the following derivation, but this method gives a very good idea about the kinds of approximations we use to solve complicated equations. Remember that in the case of tunnelling, the barrier height should be greater than the particle energy, that is $V(x) \geq E$.

Let us rewrite Equation (2.15) as follows:

$$\frac{\partial^2 \psi}{\partial x^2} = k'(x)^2 \psi \tag{2.16}$$

where

$$k'(x) = \sqrt{\frac{2m(V(x)-E)}{\hbar^2}} \tag{2.17}$$

Since $\psi(x)$ is a complex function, it may be written in terms of its amplitude $A(x)$ and phase $\phi(x)$, which are both real functions of x:

$$\psi(x) = A(x)e^{\imath\phi(x)} \tag{2.18}$$

Substituting (2.18) into (2.16), we get

$$\frac{\partial^2 A}{\partial x^2} + 2\imath \frac{\partial A}{\partial x}\frac{\partial \phi}{\partial x} + \imath A \frac{\partial^2 \phi}{\partial x^2} - A\left(\frac{\partial \phi}{\partial x}\right)^2 = k'^2 A \tag{2.19}$$

Equating the real and imaginary parts of Equation (2.19) gives us

$$\frac{\partial^2 A}{\partial x^2} - A\left(\frac{\partial \phi}{\partial x}\right)^2 = k'^2 A \tag{2.20}$$

$$2\imath \frac{\partial A}{\partial x}\frac{\partial \phi}{\partial x} + \imath A \frac{\partial^2 \phi}{\partial x^2} = 0 \tag{2.21}$$

Note that Equation (2.21) is equivalent to

$$\frac{\partial}{\partial x}\left(A^2 \frac{\partial \phi}{\partial x}\right) = 0 \tag{2.22}$$

Therefore, the solution of Equation (2.22) gives us a relation between the amplitude $A(x)$ and phase $\phi(x)$ of the wavefunction $\psi(x)$:

$$A = \frac{C}{\sqrt{|\partial \phi / \partial x|}} \tag{2.23}$$

where C is a real constant.

Till this point, the solution is exact and no approximations have been made. However, solving (2.20) without an approximation would be a challenging task. Therefore, we assume that the amplitude of the wavefunction varies slowly with x. Due to this assumption, $\partial^2 A/\partial x^2 \sim 0$, reducing Equation (2.20) to

$$\left(\frac{\partial\phi}{\partial x}\right)^2 = -k'^2 \tag{2.24}$$

The phase $\phi(x)$ can now be written as

$$\phi(x) = \pm \imath \int |k'(x)|\, dx \tag{2.25}$$

Using Equations (2.18) and (2.23) to (2.25), the wavefunction can be given as

$$\psi(x) = \frac{C}{\sqrt{k'(x)}} e^{\pm \int |k'(x)| dx} \tag{2.26}$$

A general solution for the wavefunction can be represented by a linear superposition of the positive and negative exponential terms shown in Equation (2.26).

At this point, the mathematical part of the solution is complete and we will now consider the physics involved in finding the tunnelling probability. The wavefunction $\psi(x)$ in Equation (2.26) is the solution of Schrodinger's equation within the potential barrier region. There will still be incoming, reflected and transmitted waves of the same form as seen in the section on the rectangular barrier (Section 2.1.2). Using the notations defined in that section, the wavefunction in the barrier region ψ_{II} can be written as

$$\psi_{II} = \frac{C}{\sqrt{k'(x)}} e^{\int_0^x |k'(x)| dx} + \frac{D}{\sqrt{k'(x)}} e^{-\int_0^x |k'(x)| dx} \tag{2.27}$$

As we go further into the barrier, the probability of the particle being found should decrease. Since the wavefunction is the probability amplitude (Section 1.1.2.2), this means that ψ_{II} should decrease as we go further into the barrier region. However, the first term in Equation (2.27) is an exponential that increases with x. Therefore, to make sure our wavefunction displays this behaviour, the coefficient C of the first term in the above equation should be very small. This is especially true if either the barrier width or the barrier height is very large, which is usually true in most practical situations. By neglecting the first term and imposing the boundary conditions as described in Equations (2.9) to (2.12), we get the tunnelling probability T as

$$T = \frac{|F|^2}{|A|^2} = e^{-2\gamma} \tag{2.28}$$

where

$$\frac{|F|}{|A|} \sim e^{-\int_0^L |k'(x)| dx} \tag{2.29}$$

and

$$\gamma = \int_0^L \left| \sqrt{\frac{2m(V(x) - E)}{\hbar^2}} \right| dx \tag{2.30}$$

We can now find the tunnelling probability through a general potential barrier independent of any material or band structure properties. In the following section, let us now see how the current due to tunnelling can be estimated considering the properties of the material.

2.3 Landauer's tunnelling formula

In the previous sections, we did not consider the tunnelling electrons as part of any material or band structure. Therefore, we need to connect the concepts of quantum mechanics with the band structure of the material, so that we can obtain a model for the current in a system. This is done by using the Landauer's tunnelling formula. Let us look at a simple derivation of Landauer's tunnelling formula required for developing TFET models.

All bulk crystalline materials have continuous energy bands – a continuum of available energy levels that may be occupied by an electron. The Fermi–Dirac statistics gives us this probability $f(E)$ of occupancy of a level at energy E as

$$f(E) = \frac{1}{1 + e^{(E - E_F)/kT}} \tag{2.31}$$

where k is Boltzmann's constant, E_F is the Fermi energy of the material and T is the temperature. We can observe from Equation (2.31) that a higher energy state would have a lower probability of being occupied, as electrons prefer to stay in a lower energy state. Of course, due to thermal excitations, some electrons would be found in higher energy states as well, and the number of such electrons would increase as we increase the temperature. Let us initially assume the temperature to be 0 K, so that all states below the Fermi level E_F are occupied, while all states above it are empty. Let us take a look at Figure 2.3(a), which shows two materials divided by a potential barrier. If this potential barrier were not present, electrons in one region would be free to move to the other. However, since the potential barrier is present, electrons need to tunnel through it to get to the other side.

Note that electrons would be tunnelling from both the sides, Region A to Region B and vice versa. We will first find the current $I_{A \to B}$ corresponding to electrons flowing from Region A to Region B. A similar formula will give us the current $I_{B \to A}$ corresponding to electrons flowing in the opposite direction. The net current I_{net} would be the difference between these two currents.

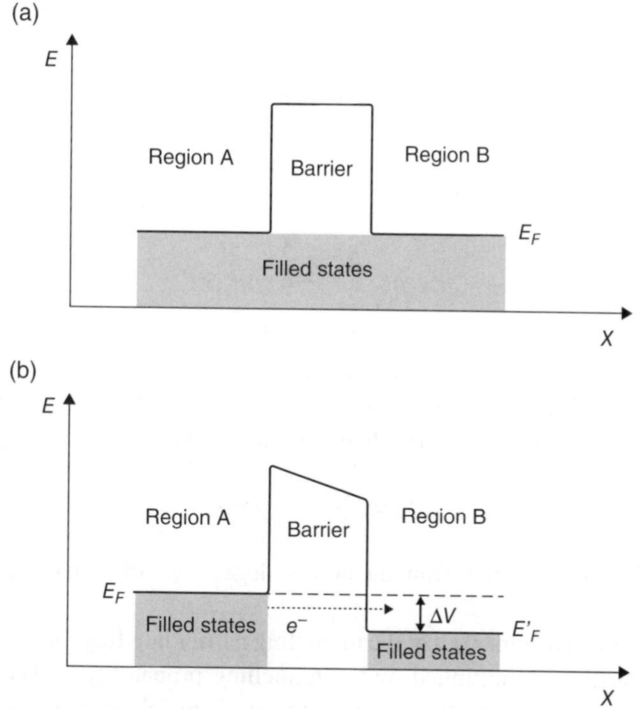

Figure 2.3 (a) Regions divided by a potential barrier with no external bias. No empty states exist in Region B for electrons to tunnel. (b) Regions divided by a potential barrier with external bias. Empty states exist in Region B, to which electrons can now tunnel.

Since electrons are fermions [2], no two electrons can have the same quantum state and an energy level can only hold two electrons – one of spin-up and one of spin-down. Therefore, movement of electrons from Region A to B (and vice versa) can take place only if there are empty states available in the other region. Initially, when no bias is applied (Figure 2.3(a)), there are no empty states on either side of the barrier. Therefore, electrons cannot tunnel through the barrier. However, when a small bias ΔV is applied, the Fermi energy of Region B is lowered by ΔV, as shown in Figure 2.3(b). The electrons having energies between $E - \Delta V$ and E can now move from Region A into the corresponding unoccupied energy levels in Region B, and thus contribute to the tunnelling current.

Let us consider Region A to be a one-dimensional potential well of length L (Section 1.2.2). Writing the wavefunction in this "potential well" in the form of two travelling waves (as shown in Equation (1.39)), we get

$$\psi_A = \frac{\imath}{2L}\left(e^{-\imath k_n x} - e^{\imath k_n x}\right) \tag{2.32}$$

where the incoming wave corresponds to $e^{ik_n x}$. Therefore, the probability density corresponding to the incoming wave is

$$\rho = 1/2L \tag{2.33}$$

For a one-dimensional potential well, the difference in the wave vector k_n between two successive states is

$$\Delta k_n = \frac{\pi}{L} \tag{2.34}$$

The total current I_{in} going into the tunnelling barrier is the summation of the individual contributions from each energy level between E and $E - \Delta V$. These individual contributions can be written by using the equation for current due to a free particle (Equation (1.28)). Their summation would, therefore, give I_{in} as

$$I_{in} = 2\sum_{E-\Delta V}^{E} qv\rho \tag{2.35}$$

where the factor of 2 comes from the double degeneracy of each energy level due to electron spin.

The current that comes out of the tunnelling barrier into Region B would be the incoming current I_{in} multiplied by the tunnelling probability T. The tunnelling probability is dependent on the energy of the incoming electron. For the nth state, having a wave vector k_n, the tunnelling probability is $T(k_n)$. Therefore, the current transmitted from Region A to Region B is:

$$I_{A\to B} = 2\sum_{E-\Delta V}^{E} T(k_n)qv\rho \tag{2.36}$$

Multiplying and dividing by π/L and using Equation (2.34) gives

$$I_{A\to B} = 2q\rho\frac{L}{\pi}\sum_{E-\Delta V}^{E} Tv\frac{\pi}{L} = 2q\rho\frac{L}{\pi}\sum_{E-\Delta V}^{E} Tv\Delta k_n \tag{2.37}$$

Substituting the value of probability density ρ from Equation (2.33),

$$I_{A\to B} = \frac{q}{\pi}\sum_{E-\Delta V}^{E} Tv\Delta k_n \tag{2.38}$$

However, as $L \to \infty, \Delta k_n \to 0$. Therefore:

$$I_{A\to B} = \frac{q}{\pi}\int_{E-\Delta V}^{E} Tv dk \tag{2.39}$$

So far we have considered the temperature to be 0 K. Let us now assume the temperature to be greater than 0 K. Due to thermal excitations, electrons can now occupy energy levels above the Fermi level E_F. The probability that an energy level is occupied by an electron is given by the Fermi–Dirac statistics (Equation (2.31)). If the probability of occupancy of each energy level in Region A is given as f_A, then the current from Region A to Region B becomes

$$I_{A\to B} = \frac{q}{\pi}\int_{E-\Delta V}^{E} Tvf_A\,dk \tag{2.40}$$

We will now use the relation for group velocity v to change the variable of integration from wave vector k to energy E:

$$v = \frac{1}{\hbar}\frac{\partial E}{\partial k} \tag{2.41}$$

This equation is similar to the group velocity in the case of waves. When we substitute Equation (2.41) into (2.40), we get the expression for the electron current going from Region A to Region B:

$$I_{A\to B} = \frac{2q}{\hbar}\int_{E-\Delta V}^{E} f_A(E)T(E)dE \tag{2.42}$$

Similarly, the current going from Region B to Region A is

$$I_{B\to A} = \frac{2q}{\hbar}\int_{E-\Delta V}^{E} f_B(E)T(E)dE \tag{2.43}$$

Equations (2.43) and (2.44) can be combined to give the final expression for Landauer's formula:

$$I_{net} = I_{A\to B} - I_{B\to A} = \frac{2q}{\hbar}\int_{E-\Delta V}^{E} (f_A(E) - f_B(E))T(E)dE \tag{2.44}$$

The above formula is extensively used in TFET models to calculate the current.

2.4 Advanced tunnelling models

In semiconductors, two different kinds of models are used to calculate the current resulting from tunnelling – local and non-local models. Non-local models treat tunnelling as a process that occurs in spatial coordinates, where electrons tunnel from one point in space to another, as shown in Figure 2.3(b). Therefore, the WKB approximation and Landauer's tunnelling formula that we have considered earlier can be classified as non-local models. Local models, on the other hand, treat tunnelling as a phenomenon taking place from one energy band to another in the *E–k* space of the material.

2.4.1 Non-local tunnelling models

Since the non-local models approach tunnelling from a spatial perspective, we need to know the spatial dependence of all the parameters in Schrodinger's equation to obtain the tunnelling probability. If an external bias is applied to a semiconductor, the shape of the potential barrier $V(x)$ is complicated. The band structure of the semiconductor also needs to be incorporated into the potential $V(x)$ in Schrodinger's equation. Moreover, the potential in the semiconductor depends on the current (Ohm's law), while the current in turn depends on the shape of the potential barrier (Schrodinger's equation). All these complicated spatial dependencies make it difficult to solve Schrodinger's equation analytically. Therefore, numerical approaches need to be applied to obtain a solution for calculating the tunnelling probability. However, because all these complex dependencies are incorporated, non-local models provide an accurate estimation of the current in the device although they are difficult to solve analytically. As non-local models cannot be used to obtain analytical models, they are used in device simulation. On the other hand, analytical models are essential for use in circuit simulation and to understand the functioning of devices. To develop analytical models, we use local tunnelling models.

2.4.2 Local tunnelling models

To understand local models, let us consider the problem from an energy band perspective. Semiconductors have various energy bands, which result from the interaction of the individual fields created by the atoms in its lattice. These energy bands are usually represented in an *E–k* diagram, which shows the allowed values of energy E that the electron can take. The *E–k* diagram of an intrinsic direct bandgap semiconductor is shown in Figure 2.4(a). We can observe the presence of a valence band and a conduction band, separated by the bandgap – a potential barrier. Since we have considered an intrinsic material, there are very few electrons in the conduction band. In the presence of an external electric field, these few electrons will flow to generate a very low current. However, if this electric field is sufficiently large, it is possible for electrons to tunnel from the valence band to the conduction band without a change in energy, as shown in Figure 2.4(b). In other words, the electric field causes the bands in the semiconductor to be modified in such a way that the electrons in the valence band can tunnel through the potential barrier (the bandgap) and reach the conduction band. Now that there are sufficient electrons in the conduction band due to tunnelling, an appreciable current can flow in the material.

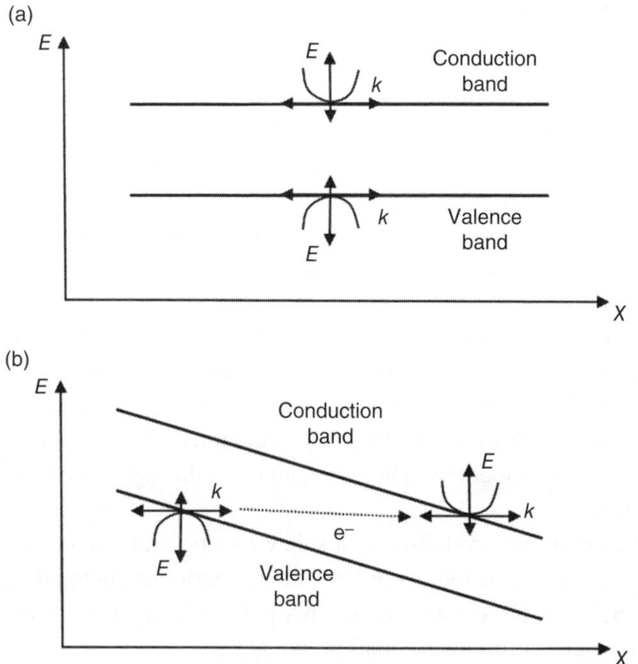

Figure 2.4 (a) Energy bands of a semiconductor under zero bias, with inset E–k diagrams. (b) Energy bands of a semiconductor with an electric field, with inset E–k diagrams. Non-local occurs, with the electron transitioning from the valence band maxima to the conduction band minima in the E–k diagram.

An important approximation made while deriving local models is that the electric field is assumed to be constant, which is rarely true in a real device. Because of the assumption of a constant electric field, local models predict a single tunnelling rate throughout the device. In case the electric field is position-dependent, we estimate the tunnelling rate using the local electric field at every point in the device. The tunnelling rate at each point is then integrated throughout the device to give the total number of electrons that have tunnelled from the valence band to the conduction band. Since no generation or recombination is assumed, the total current is dependent only on the rate at which the electrons tunnel from the valence band to the conduction band. Therefore, in local models, it is assumed that each electron that reaches the conduction band is swept away to form a part of the total current. Local models are extensively used in analytical modelling of TFETs as they give an analytical expression for the tunnelling rate at each point.

We will now discuss two commonly used local models for calculating the tunnelling rate in semiconductors.

2.4.2.1 Kane's model

Developed by E.O. Kane in 1959 [3], this is one of the oldest and most widely used models for calculating the band-to-band tunnelling rate in TFET models. The derivation of Kane's model is very complicated. However, a brief summary of the procedure followed in the paper is presented below.

As mentioned in the preceding section, we will be approaching tunnelling as a phenomenon occurring in the *E–k* diagram rather than in the spatial coordinates. We would, therefore, need to reframe Schrodinger's equation accordingly. In this new representation of the Schrodinger equation, there must be no spatial terms. The terms that depend on space in Schrodinger's equation are: (i) the derivatives that are with respect to spatial coordinates and (ii) the spatial dependence of the external potential $V(x)$. These spatial terms must now be reframed as functions of either energy E or momentum k. This reformulation of Schrodinger's equation is done by using Bloch functions. Therefore, let us go through a brief introduction to Bloch functions.

Semiconductors are crystalline solids, having a periodic structure. Due to such a structure, the atoms of the crystal produce a periodic potential V_{per} with the same periodicity as the crystal. When this potential V_{per} is substituted into the time-independent Schrodinger equation, we get

$$-\frac{\hbar^2}{2m}\frac{\partial^2\psi}{\partial x^2} + V_{per}\psi = E_n\psi \tag{2.45}$$

Note that we have added a subscript n to the energy, since different energy bands in the crystal correspond to different energies. In solid state physics, the solutions to the above Equation (2.45) are known as Bloch functions$\psi_{n,k}$, which have the form:

$$\psi_{n,k} = e^{ikx}u_{n,k}(x) \tag{2.46}$$

where $u_{n,k}(x)$ is a periodic function with the same periodicity as the crystal. The subscript n indicates that this Bloch function belongs to the nth energy band (i.e. its energy is E_n) and $k$denotes the *crystal momentum* of this wavefunction. A general wavefunction Ψ can, therefore, be written as

$$\Psi = \Sigma a_n(k)\psi_{n,k} \tag{2.47}$$

where $a_n(k)$ are complex coefficients, as described in Equation (1.10). With this brief idea of Bloch functions, let us proceed with the derivation of Kane's model.

First we will consider the familiar Schrodinger equation in spatial coordinates for our particular physical situation, which is a crystal lattice (the semiconductor) to which an external bias is applied. The potential term $V(x)$ in Schrodinger's equation will, therefore, be a sum of the contribution due to the crystal lattice (V_{per}) and the contribution due to the external bias (V_{ext}). It is assumed that the external bias V_{ext} leads to a uniform electric field E_{ext} in the crystal lattice. Since the potential due to a uniform electric field is linear ($V_{ext} = -qE_{ext}x$), we can write the time-independent Schrodinger equation (1.17) as

$$-\frac{\hbar^2}{2m}\frac{\partial^2\psi}{\partial x^2} + \left(V_{per} - qE_{ext}x\right)\psi = E\psi \tag{2.48}$$

We will now use the Bloch functions to convert Schrodinger's equation from this spatial form to what is known as the "crystal momentum representation" [4]. The crystal momentum representation is simply another form of Schrodinger's equation without any spatial dependence. In this representation, every operator and wavefunction in Schrodinger's equation is written as a linear combination of the Bloch functions. For decomposing the wavefunction ψ into a linear combination of Bloch functions, we only need to substitute the equation for a general wavefunction (2.47) into Schrodinger's equation, which will give

$$\Sigma a_n(k)\left(\frac{\hbar^2}{2m}\frac{\partial^2\psi_{n,k}}{\partial x^2} + V_{per}\psi_{n,k}\right) - \Sigma a_n(k)qE_{ext}x\psi_{n,k} = \Sigma a_n(k)E\psi_{n,k} \tag{2.49}$$

By comparing with Equation (2.45), we can observe that the term in the brackets is the energy of the nth band E_n. Making this substitution, and writing the equation for a single band, i.e. a single $a_n(k)$, we get

$$a_n(k)E_n - a_n(k)qE_{ext}x = a_n(k)E \tag{2.50}$$

Note that there are two terms corresponding to energy in the above equation – E_n corresponding to the energy solely due to the crystal and E corresponding to the total energy of the electron. We have now decomposed the wavefunction in terms of Bloch functions, leaving the position operator x as the only spatially dependent operator in Equation (2.50). Therefore, to make Schrodinger's equation completely spatially independent, we now need to decompose this operator x in terms of energy E and momentum k. This procedure involves advanced quantum mechanics, and only the final result will be provided here.

The position operator x in the crystal momentum representation is decomposed into an "intraband" operator $\imath\partial/\partial k$ and an "interband" operator $X_{nn'}(k)$, where n represents one energy band and n' represents another energy band. The Schrodinger equation in the crystal momentum representation can, therefore, be written as

$$\left[E_n(k) - \imath qE_{ext}\frac{\partial}{\partial k} - E\right]a_n(k) - \sum\nolimits_{n'} qE_{ext}X_{nn'}a_{n'}(k) = 0 \tag{2.51}$$

The position of any electron is found to depend on two factors – the contribution of the energy band that it belongs to and the contribution due to the interaction of that energy band with all the other energy bands in the crystal lattice. The intraband operator $\imath\partial/\partial k$ corresponds to the spatial behaviour of the electron resulting from its own energy band. The interband operator $X_{nn'}(k)$, on the other hand, corresponds to the spatial behaviour of the electron due to the contribution of all the other energy bands inside the crystal lattice. It is this interband term that couples the various bands within the crystal, leading to tunnelling. In case no external field were present, there would be no position operator in Equation (2.49) and there would be no interband terms in the final equation to connect the different bands. Thus, electrons would be confined to their bands and there would be no.On the other hand, as the field is increased, the coupling term $\sum_{n'} qE_{ext}X_{nn'}a_{n'}(k)$ increases, leading to greater tunnelling.

The crystal momentum representation of Schrodinger's equation is difficult to solve. Therefore, as a first-order approximation, the eigenfunctions of Equation (2.51) are written after ignoring the interband terms $\sum_{n'} qE_{ext}$ $X_{nn'}a_{n'}(k)$. This step is able to simplify the problem by including the effect of the electric field on the bands themselves, without including the effects of interaction between various bands. These eigenfunctions are then used, along with the interband operator $X_{nn'}$, to calculate the rate of tunnelling from band n to band n'.

It should be noted that the above derivation was for a general periodic potential V_{per}. Kane's model can now be derived for a semiconductor by modelling the periodic potential V_{per} as two bands interacting via the $k{\cdot}p$ perturbation [5]. The final result for the band-to-band generation rate G_{btb} is given as

$$G_{btb} = \frac{E_{ext}^2 m_r^{1/2}}{18\pi\hbar^2 E_G^{1/2}} \exp\left\{\frac{-\pi m_r^{1/2} E_G^{3/2}}{2\hbar|E_{ext}|}\right\} = A\frac{E_{ext}^2}{E_G^{1/2}} \exp\left\{-B\frac{E_G^{3/2}}{|E_{ext}|}\right\} \tag{2.52}$$

where m_r is the reduced mass of the charge carrier.

The above model (2.52) is for a direct semiconductor, where the electron tunnels from the valence band to the conduction band without any change in its momentum. In the case of an indirect semiconductor, the electron must change its momentum when tunnelling from the valence band edge to the conduction band edge. To ensure conservation of momentum, this change in momentum is transferred to the semiconductor lattice in the form of quantised lattice vibrations (known as "phonons"), and the final result in this case is [6]:

$$G_{btb} = \frac{E_{ext}^{5/2} m_r^{1/2}}{18\pi\hbar^2 E_G^{1/2}} \exp\left\{\frac{-\pi m_r^{1/2} E_G^{3/2}}{2\hbar|E_{ext}|}\right\} = A\frac{E_{ext}^{5/2}}{E_G^{1/2}} \exp\left\{-B\frac{E_G^{3/2}}{|E_{ext}|}\right\} \tag{2.53}$$

Note that the difference between the band-to-band generation rate in direct and indirect bandgap semiconductors is only in the power of the external electric field E_{ext} in the pre-exponential term. In the case of direct semiconductors, this power is 2 (i.e. E_{ext}^2), while in the case of indirect semiconductors, this power is 2.5 (i.e. $E_{ext}^{5/2}$).

While the Kane's model is used extensively to calculate the band-to-band tunnelling rate, there are many effects that this model ignores, such as the effects of traps and density of states. Many tunnelling models have subsequently been developed that use Kane's model as a base, and are more accurate in calculating the band-to-band tunnelling rate. One such model is the Hurkx model, which includes the effects of traps and density of states while calculating the band-to-band tunnelling rate.

2.4.2.2 Hurkx model

This model was developed by Hurkx in 1992 [7] as a recombination model that included the effects of tunnelling. However, the Hurkx model is also used as a tunnelling model by many device simulators. The advantage of the Hurkx model over Kane's model is the inclusion of effects of trap-assisted tunnelling, and density of states. A qualitative outline of the model's derivation is now presented.

The net recombination rate R in the Hurkx model is taken to be the sum of recombination rates due to traps (including both SRH recombination and trap-assisted tunnelling) R_{trap} and due to band-to -band tunnelling R_{bbt}:

$$R = R_{trap} + R_{bbt} \tag{2.54}$$

The expression for R_{trap} is given as

$$R_{trap} = N_T \frac{c_n c_p n_t p_t - e_n e_p}{c_n n_t + c_p p_t + e_n + e_p} \tag{2.55}$$

where N_T is the trap density and n_t and p_t are the densities of electrons and holes having capture rates c_n and c_p, respectively. The probability per unit time for emission of electrons and holes is e_n and e_p, respectively. You may notice that this expression is very similar to the one given for SRH recombination [8]. The reason for this is that both the derivations follow the same procedure. In the Hurkx model the R_{trap} term incorporates the effects of tunnelling while calculating the density of carriers (n_t and p_t) and their emission probabilities per unit time (e_n and e_p) since these quantities are increased due to tunnelling.

Let us first consider the effect of tunnelling on the carrier densities n_t and p_t. In case high electric fields are present, electrons can tunnel into the bandgap and get captured by traps, as shown in Figure 2.5(a). Due to this, the density of carriers

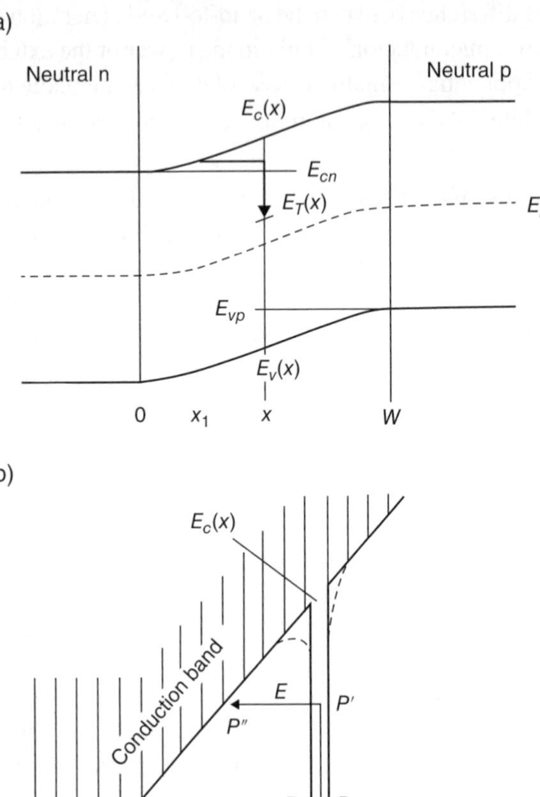

Figure 2.5 (a) Electrons from the conduction band of the n-region can tunnel traps having energy lower than the conduction band minima. (b) Electrons occupying traps having energy above the conduction band minima can tunnel to the conduction band [7]. Source: Reproduced with permission of IEEE.

within the depletion region increases. The expression for the carrier densities is obtained by solving the effective-mass Schrodinger equation for a linear potential. The electron carrier density $n_t(x)$ is

$$n_t(x) = n(x) + \int_{\delta}^{x} \left(-\frac{dn(x)}{dx} \right)_{x=x_1} \frac{Ai^2[\gamma(x-x_1)]}{Ai^2[0]} dx_1 \tag{2.56}$$

where Ai is the Airy function, $\gamma = \left(2q\bar{F}m^*\hbar^{-2}\right)^{1/3}$, $\bar{F}$ is the average electric field and $m*$ is the effective mass. The first term on the right-hand side $n(x)$ is the density of electrons without considering the effects of tunnelling. The second term is the contribution due to an electron at a position x_1 tunnelling to a trap at x. The integration is performed over all the possible locations x_1 from where an electron can tunnel into a trap at x. A similar expression can be written for the hole density.

Let us now look at how tunnelling affects the emission rates. In case tunnelling is not considered, an electron in a trap needs thermal excitation to cover the entire trap depth E_C–E_T to reach the conduction band. However, an electron can be partially excited through only a part of the trap depth, and can then tunnel through the remaining barrier, as shown in Figure 2.5(b). The emission probability can, therefore, be written as

$$e_n = e_{n0}\left[1+\frac{1}{kT}\int_0^{\Delta E_n} e^{E/kT}\frac{Ai^2\left(2m^*\gamma^{-2}\hbar^{-2}E\right)}{Ai^2(0)}dE\right] \tag{2.57}$$

where e_{n0} is the emission probability if tunnelling is not considered.

In case the potential cannot be approximated as linear, the tunnelling probability obtained for a constant electric field $Ai^2[\gamma(x-x_1)]/Ai^2(0)$ in Equations (2.56) and (2.57) can be replaced by the tunnelling probability obtained from the WKB approximation (Equation (2.30)). This completes the calculation for the trap related recombination rate.

To calculate the recombination rate due to band-to-band tunnelling, Hurkx has used a slightly modified form of Kane's formula:

$$R_{bbt} = -B|F|^{\sigma}De^{-F_0/|F|} \tag{2.58}$$

where F is the external field at the given point, $\sigma=2$ for direct transitions and $\sigma=5/2$ for indirect transitions. The pre-factor B is a constant, while F_0 is proportional to $E_g^{3/2}$, where E_g is the bandgap. The term D was not present in Kane's model and is a function that accounts for the relative position of the Fermi levels E_{fp} and E_{fn} in the neutral p- and n-regions, respectively:

$$D\left(V,E_{fn},E_{fp}\right) = \frac{1}{\exp\left[\left(-E_{fp}-qV\right)/kT\right]+1} - \frac{1}{\exp\left[\left(-E_{fn}-qV\right)/kT\right]+1} \tag{2.59}$$

where V is the electric potential in the region. To calculate the effect of density of states, it is assumed that an electron from the p-region (point A in Figure 2.6) tunnels to the n-region (point B in Figure 2.6), both at a potential energy $E=qV$. The term D is, therefore, the difference between the probability of occupation at point A and point B. Outside the depletion region, on the neutral n-side, the function D is almost zero as the potential energy qV equals the neutral n-side Fermi energy E_{fn}. Physically, this signifies that there are no final states in the p-region at this energy into which an electron can tunnel. Similarly, this function is negligible outside the depletion region on the neutral

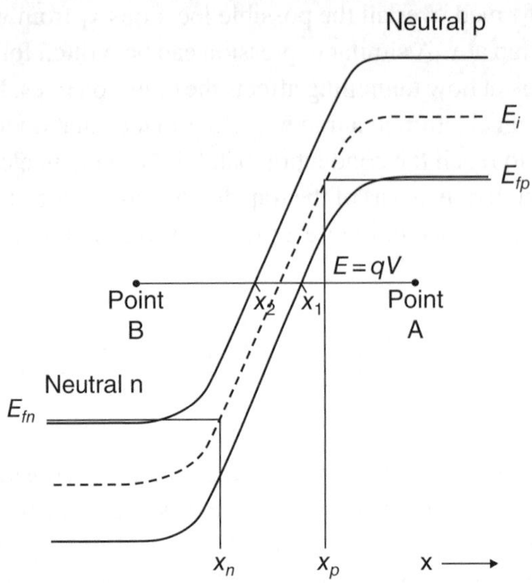

Figure 2.6 Incorporation of density of states in the Hurkx model [7]. Source: Reproduced with permission of IEEE.

p-side, as there are no initial states from which an electron can tunnel. Therefore, this function D is able to capture the effect of density of states on band-to-band tunnelling.

References

[1] D. J. Griffiths, *Introduction to Quantum Mechanics*, 2nd edn, Pearson Education, 2005.

[2] R. Eisberg and R. Resnick, *Quantum Physics of Atoms, Molecules, Solids, Nucleii, and Particles*, 2nd edn, John Wiley & Sons, 2010.

[3] E. O. Kane, "Zener Tunneling in Semiconductors", *J. Phys. Chem. Solids*, vol. 12, no. 2, pp. 181–188, January 1960.

[4] E. N. Adams II, "The Crystal Momentum as a Quantum Mechanical Operator", *J. Chem. Physics*, vol. 21, no. 2013, 1953.

[5] C. Kittel, *Quantum Theory of Solids*, 2nd edn, John Wiley & Sons, 1987.

[6] E. O. Kane, "Theory of Tunnelling", *Journal of Applied Physics*, vol. 32, no. 83, 1961.

[7] G. A. M. Hurkx, D. B. M. Klaassen and M. P. G. Knuvers, "A New Recombination Model for Device Simulation Including Tunnelling", *IEEE Trans. Electron Devs.*, vol. 39, no. 2, February 1992.

[8] S. Sze, *Physics of Semiconductor Devices*, John Wiley & Sons, Inc., New York, 1981.

3

The tunnel FET

3.1 Device structure

We will now familiarise ourselves with the basic structure of the device we will be modelling in this book – the tunnel FET. Understanding the structure of any device goes beyond simply knowing the different regions, materials and dopings used. It is essential to understand the reason behind the particular structure chosen, and how each parameter in the structure optimises some desirable characteristics in the device behaviour. Therefore, in this chapter, apart from introducing the structure of the TFET, we will have a brief qualitative discussion about the behaviour of the TFET.

3.1.1 The need for tunnel FETs

An essential area of focus in semiconductor technology, and devices in particular, is the continuous scaling down of the device dimensions. We have come a long way in this pursuit – from gigantic vacuum tubes that led to computers the size of large rooms, to the first MOSFET with a gate length of 300 μm, to transistors with gate lengths of 14 nm or less. This has led to integrated circuits containing billions of transistors.

Scaling the length of a MOSFET has many benefits, besides the increased number of transistors in a chip. A reduced gate length leads to a reduced gate capacitance, thereby increasing the switching speed of the circuit. Moreover, the voltage scaling that is a necessary part of device miniaturisation also causes reduction in the power consumption of the device.

Tunnel Field-Effect Transistors (TFET): Modelling and Simulation, First Edition. Jagadesh Kumar Mamidala, Rajat Vishnoi and Pratyush Pandey.

However, as the device dimensions were reduced to 50 nm and the power supply to 0.5 V, the OFF-state power consumption of MOSFETs became a major challenge. The drain current of a MOSFET is controlled by the thermionic emission from the source into the channel. As the gate voltage increases, the potential barrier between the source and the channel decreases, leading to an increase in the drain current. This leads to two problems – a larger OFF-state current due to subthreshold conduction and a higher subthreshold slope. The subthreshold slope (SS) of a MOSFET is the change in gate voltage V_G required to increase the drain current I_{DS} by a factor of 10. A lower subthreshold slope would allow for a higher ratio of ON-current to OFF-current (I_{ON}/I_{OFF}), and would lead to a lower power dissipation in the OFF-state. The subthreshold slope of a MOSFET is given as

$$SS = \left(\frac{d(\log_{10} I_{DS})}{dV_G}\right)^{-1} = 2.3\frac{kT}{q}\left(1 + \frac{C_d}{C_{ox}}\right) \tag{3.1}$$

where C_d is the depletion region capacitance and C_{ox} is the gate oxide capacitance. As we can see from Equation (3.1), the lowest possible value of the subthreshold slope for a MOSFET is 2.3 kT/q, which is 60 mV/decade at room temperature. For example, to get an $I_{ON}/I_{OFF} = 10^5$, we need to apply a gate voltage of $5 \times 60\ \text{mV} = 0.3\,\text{V}$. Therefore, it is difficult to scale down the supply voltage if we want to realise a large I_{ON}/I_{OFF} ratio. To overcome these fundamental limitations of a larger OFF-state current and poor subthreshold slope of the MOSFET, an alternate device was necessary.

One such device is the tunnelling field-effect transistor (TFET) [1–6]. TFETs can exhibit subthreshold slopes below 60 mV/decade due to a fundamental difference in the mechanism of current control as compared to MOSFETs. In MOSFETs, the current depends on the thermionic emission of free carriers across the potential barrier between the source and the channel. On the other hand, the current in TFETs depends on the charge carriers tunnelling through a potential barrier between the source valence band and the channel conduction band. As this potential barrier is very wide in the OFF-state of the device, TFETs exhibit very low OFF-state current.

Apart from the limitation imposed by the subthreshold slope, MOSFETs in the sub-50 nm channel regime also suffer from various short channel effects, such as drain-induced barrier lowering, threshold voltage roll-off, charge sharing between gate and drain, etc. [7]. As we shall study later in the chapter, TFETs have a greater immunity to these short channel effects.

It may be pointed out that TFETs differ from the MOSFET only in the type of source doping. Therefore, the integration of the TFET fabrication process with the current MOSFET fabrication process would be easy.

Let us now examine the structure of this device.

3.1.2 Basic TFET structure

We will now examine the basic structure of a tunnelling FET. Later, we will study many variations of this structure, but the working principle of the TFET is based on this basic arrangement of regions, doping and terminals.

Figure 3.1(a) shows the basic structure of an n-channel TFET. The device has three regions – the source, the channel and the drain. Comparing the structure of an n-channel TFET with that of an n-channel MOSFET (Figure 3.1(b)), we find that the source doping in a TFET is p-type, whereas it is n-type in the MOSFET. This is the only major difference between a TFET and a MOSFET. The channel region

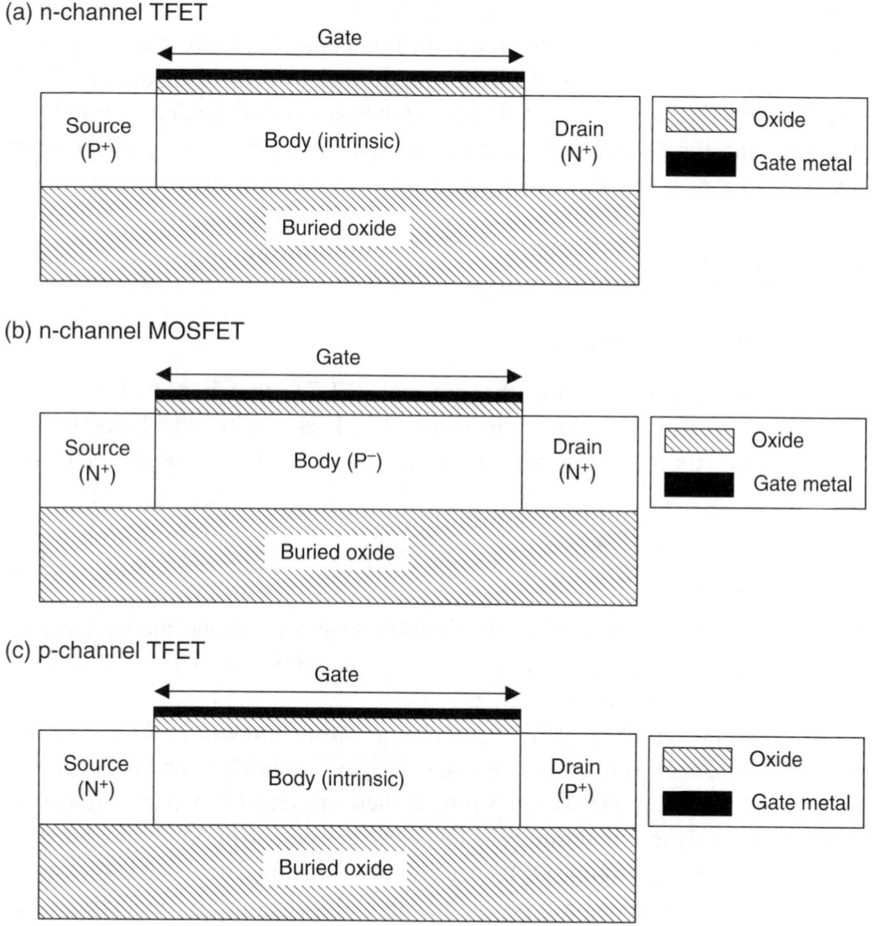

Figure 3.1 Basic structure of a TFET.

in the TFET is usually intrinsic, or very lightly doped. We will now qualitatively examine the behaviour of TFETs.

3.2 Qualitative behaviour

The behaviour of any transistor is usually described by its current characteristics – a plot of the current flowing in the device under various biasing conditions. For a transistor, the most significant current characteristics are the transfer characteristics and the output characteristics. These are plots of the drain current in the transistor with respect to the gate and the drain bias, respectively.

In this section, we will develop a qualitative understanding of the behaviour of a TFET. The effect of biasing on the drain current of a TFET is best understood by observing the band diagrams of a TFET under various biasing conditions. While studying these, we will qualitatively predict the variation of the drain current with changing bias. Finally, we will compare our qualitative predictions with the actual characteristics of a TFET.

3.2.1 Band diagram

3.2.1.1 Thermal equilibrium

The thermal equilibrium band diagram of a TFET, that is with no external bias ($V_G = V_S = V_D = 0$), is shown in Figure 3.2. There are two depletion regions formed – one at the source–channel junction and the other at the channel–drain junction.

3.2.1.2 OFF-state

The TFET is in OFF-state when the drain voltage $V_{DS} > 0$ and the gate voltage $V_{GS} = 0$, which is similar to the OFF-state of a MOSFET. The band diagram for this case is shown in Figure 3.3.

In the OFF-state of the TFET, any charge carriers present in the conduction band of the channel would have a tendency to drift to the drain and thus generate a current. However, as the source is p-type, there are very few free electrons in its conduction band, and therefore very few electrons can be injected into the channel. This leads to a negligible OFF-state current. We should note that in the case of a MOSFET, the source is n-type and has free electrons in its conduction band. Through thermionic emission, a few of these electrons will be injected into the channel over the potential barrier at the source–channel junction. This leads to a higher OFF-state current in a MOSFET as compared to a TFET.

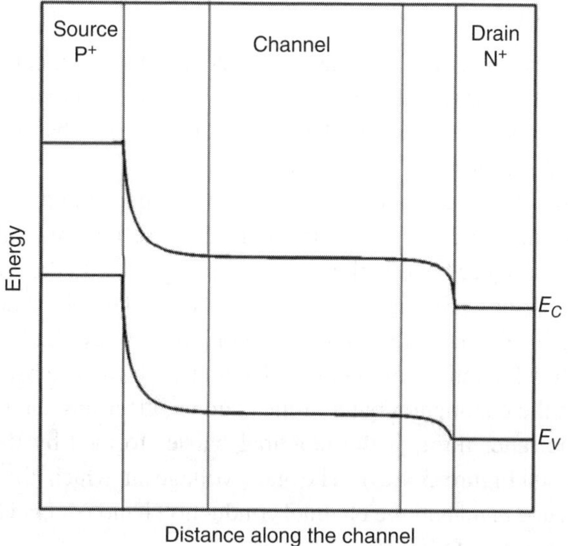

Figure 3.2 Band diagram at the surface of a n-channel TFET in thermal equilibrium (i.e. at zero bias).

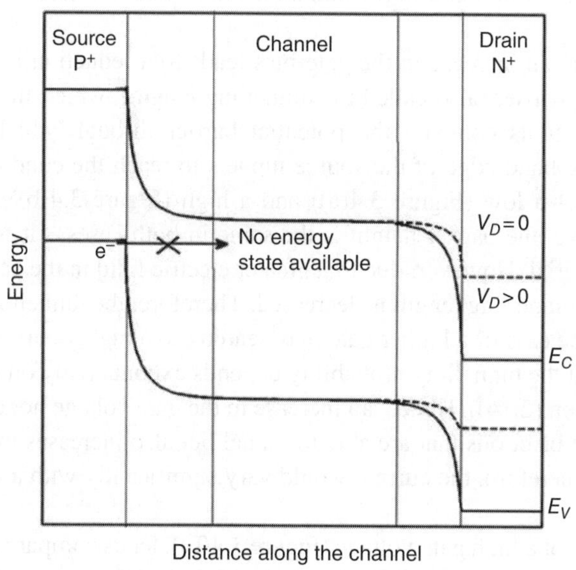

Figure 3.3 Band diagram along the surface of a TFET in the OFF-state (i.e. at V_{GS} = 0 V).

3.2.1.3 ON-state

For current to flow in the device, charge carriers need to be injected into the conduction band of the channel. In the case of TFETs, since there are negligible free electrons in the conduction band of the p-type source, these charge carriers originate from the valence band of the source.

As we increase the gate voltage V_{GS}, the energy bands in the channel change with respect to the source as shown in Figure 3.4. At a certain value of the gate voltage V_{GS}, the valence band in the source gets aligned with the conduction band in the channel, as shown in Figure 3.4(a). In the OFF-state of the device, the electrons in the valence band of the source did not have any available energy state in the channel into which they could tunnel. Now that the valence band of the source is aligned with the conduction band of the channel, electrons can tunnel from the former into the latter through the potential barrier formed by the bandgap E_G (dashed shape in Figure 3.4(a)). The gate voltage at which this alignment of the source valence band and the channel conduction band occurs is the beginning of the ON-state of the TFET.

Notice that in Figure 3.4(a) the valence band of the source and the conduction band of the channel are just aligned. As the gate bias is further increased (Figure 3.4(b)), the bands in the channel region are further lowered in energy, and electrons occupying energy levels from the valence band edge of the source $E_{V,Source}$ to the conduction band edge of the channel $E_{C,Channel}$ can tunnel to the conduction band in the channel. This leads to a steep increase in the current.

In addition, an increase in the gate bias leads to a reduction in the length of the tunnelling barrier (also called the tunnelling length), which further increases the current. Let us compare the potential barrier through which an electron on the valence band edge of the source tunnels to reach the conduction band in the channel for a low (Figure 3.4(a)) and a high (Figure 3.4(b)) gate bias. As we can observe, the barrier height is the same in both cases – it is the bandgap E_G of the material. However, due to a greater electric field in the case of a higher gate bias, the tunnelling length is decreased. Therefore, the tunnelling probability increases in the case of a higher gate bias, leading to a higher current. We should remember that the tunnelling probability depends exponentially on the tunnelling width (Equation (2.14)). Hence, an increase in the gate voltage not only increases the number of electrons that are able to tunnel but also increases their tunnelling probability. Therefore, the current would vary significantly with a changing gate voltage.

In the case of a high gate voltage (Figure 3.4(b)), let us compare the tunnelling probability of an electron at the valence band edge of the source ($E_{V,Source}$) with that of an electron occupying the lowest energy from which tunnelling can occur

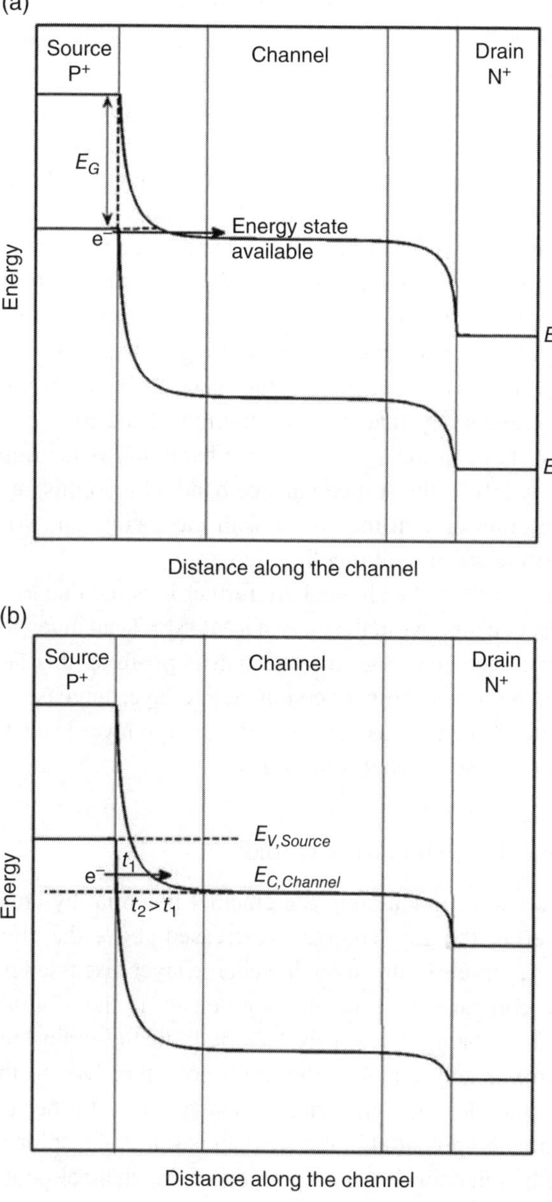

Figure 3.4 Band diagram along the surface of a TFET (a) at the beginning of the ON-state and (b) deep into the ON-state.

($E_{C,Channel}$). The tunnelling probability depends on the height and the length of the potential barrier. The height of the potential barrier is the same for all energy levels – it is the bandgap of the material. The length of the potential barrier depends on the slope of the energy bands (i.e. the electric field) – a higher slope leading to a shorter tunnelling length. Using our knowledge of the p–n junction diode, the slope of the energy bands is highest at the junction of p-type and n-type regions. In the case of TFETs, this junction lies almost at the edge of the source, since the source is heavily doped as compared to the channel. Hence, the electron at the source edge occupying the energy level $E_{V,Source}$ tunnels through a potential barrier having a shorter tunnelling length (L_{T_1}) as compared to the electron at energy $E_{C,Channel}$ (L_{T_2}). Therefore, the tunnelling probability of the electron at energy $E_{C,Channel}$ is lower than that of the electron at the valence band edge $E_{V,Source}$. Thus, we can say that the contribution to the total current decreases for lower energy states in the source valence band and is negligible for energy states that are very low in the source valence band. Due to this, at a high enough gate voltage, the drain current increment with increasing gate voltage is almost solely due to a decrease in the tunnelling length.

As the energy bands in the channel are further lowered due to increasing gate bias, another effect occurs. As in the case of a MOSFET, an inversion charge layer starts forming in the channel. As the gate bias is positive, this layer consists of electrons. The magnitude of this inversion charge layer increases with gate bias, and at sufficiently high gate bias the inversion charge layer leads to "pinning" of the channel potential, as we shall now study.

3.2.1.4 Pinning of the channel potential

Till this point, we were modulating the channel potential by changing the gate voltage. However, as the gate voltage is increased above the drain voltage (i.e. $V_{GS} > V_{DS}$), the magnitude of the inversion charge layer discussed in the preceding section becomes comparable to the electron density in the n^+-drain region. This leads to the channel being effectively "shorted" to the drain, and the channel potential is approximately equal to the drain potential. Due to this, the energy bands in the channel do not significantly change with a further increase in gate voltage. Therefore, the potential in the channel is said to be "pinned" to the drain potential. This phenomenon is called pinning of the channel potential. It should be noted that this pinning does not occur precisely when $V_{GS} = V_{DS}$, since the magnitude of the inversion charge changes continuously, due to which the gate control of the channel potential is reduced when $V_{GS} \sim V_{DS}$ and is completely lost when $V_{GS} > V_{DS}$.

In the preceding section, we discussed that an increase in the gate bias increases the current in the device due to two factors. The first was an increase

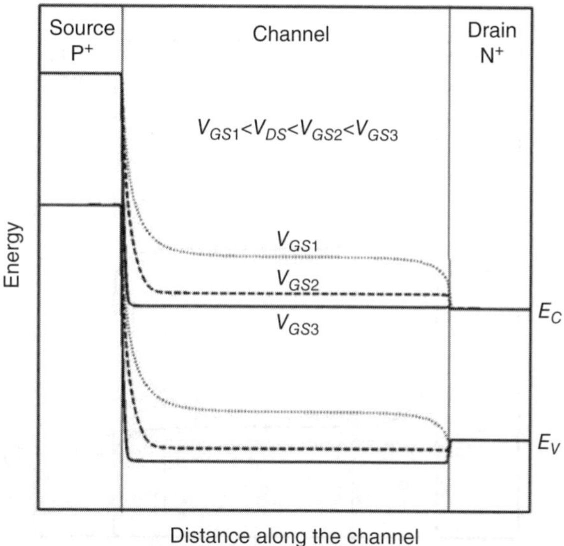

Figure 3.5 Band diagram along the surface of the TFET with increasing value of the gate voltage (V_{GS}) in the ON-state.

in the number of energy levels in the source valence band from where tunnelling is possible, which occurred because the conduction band in the channel was further lowered in energy. However, when $V_{GS} > V_{DS}$, the channel potential is pinned to the drain potential, and hence an increase in gate bias will no longer lead to a lowering of the conduction band energy in the channel.

The other factor contributing to a higher current with an increase in gate bias was the increased electric field in the source–channel junction, which led to a reduction in the tunnelling length. As we can observe from a comparison of the energy band diagrams in the source–channel junction for three gate biases $V_{G3} > V_{G2} > V_D > V_{G1}$ (Figure 3.5), the electric field increases with increasing gate bias even when the channel potential is pinned to the drain potential.

Therefore, when $V_{GS} > V_{DS}$, the current in the device increases with gate bias, but at a lower rate than in the case of $V_{GS} < V_{DS}$.

3.2.1.5 Ambipolar behaviour

In the previous sections, a positive gate voltage was applied with the drain voltage $V_{DS} > 0$. Let us now observe the behaviour of a TFET when a negative gate voltage is applied.

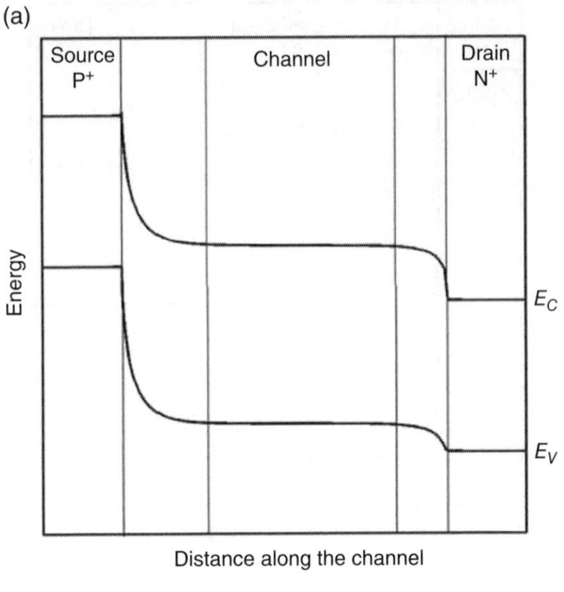

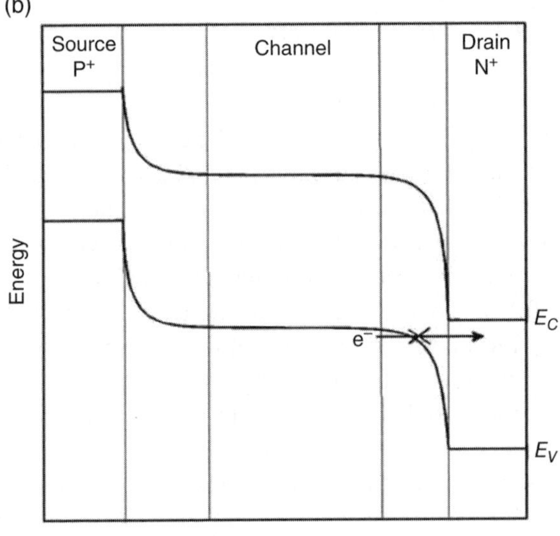

Figure 3.6 *Band diagram along the surface of a TFET (a) for a zero gate voltage (V_{GS}) and a positive drain voltage (V_{DS}), (b) for a negative gate voltage (V_{GS}) and a positive drain voltage (V_{DS}), (c) for a negative gate voltage (V_{GS}) such that the valence band in the source gets aligned with the conduction band in the drain and a positive drain voltage (V_{DS}), (d) for a more negative gate voltage (V_{GS}) as compared to Figure 3.6 (c) and a positive drain voltage (V_{DS}) and (e) with an increasing negative value of the gate voltage (V_{GS}) and a positive drain voltage (V_{DS}).*

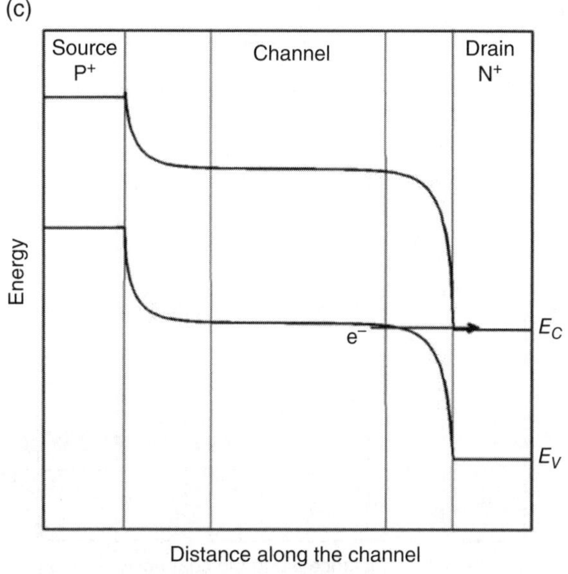

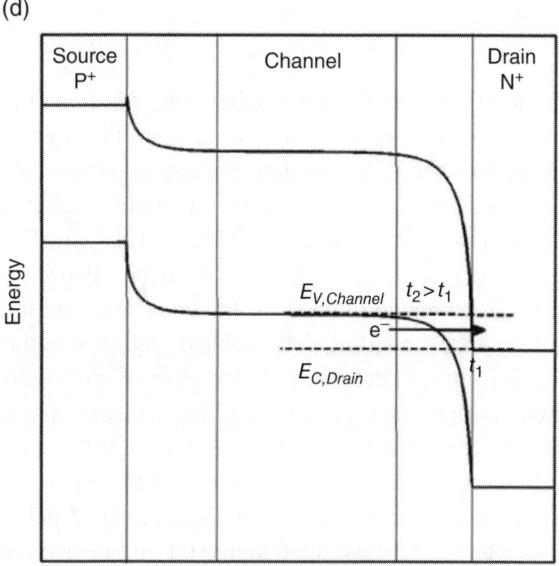

Figure 3.6 (Continued)

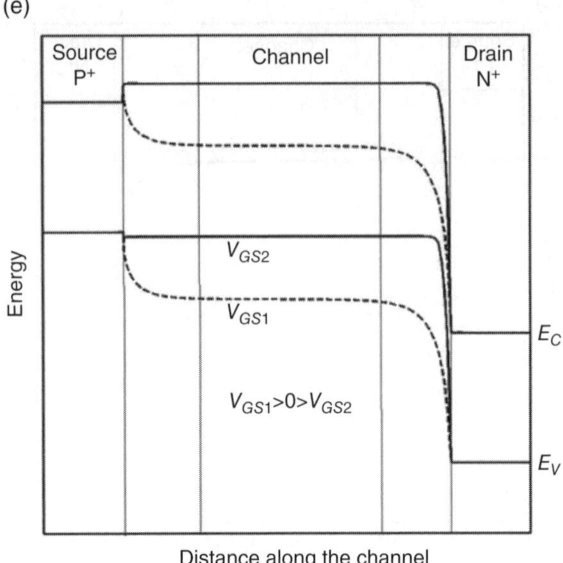

Figure 3.6 (Continued)

Initially, as $V_{GS} = 0$ (Figure 3.6(a)), no electrons can tunnel from the source valence band to the channel conduction band or from the channel valence band to the drain conduction band. As the gate voltage is decreased below 0 V, the energy band of the channel moves upward with respect to the source (Figure 3.6(b)). When the valence band of the channel is aligned with the conduction band of the drain (Figure 3.6(c)), electrons from the valence band of the channel can tunnel into the conduction band of the drain, resulting in a current flow. Note that the electrons are tunnelling in the same direction as in the case of a positive gate bias, that is from the left to the right (compare Figure 3.4(b) and (c)). This results in the device current having the same polarity even at a negative gate bias. A further increase in the negative gate bias (Figure 3.6(d)) causes a significant increase in the drain current due to (i) reduction of the tunnelling length at the channel–drain junction and (ii) an increase in the number of states in the channel valence band from where electrons can tunnel to the conduction band of the drain. This is called ambipolar conduction.

As the negative gate bias is further increased, the channel potential gets pinned to the source potential (Figure 3.6(e)). Therefore, as discussed in the section on pinning (Section 3.2.1.4), the channel potential no longer depends on the gate potential. After this point, increasing negative gate voltage causes an increase in the drain current solely due to a decrease in the tunnelling length. Therefore, the drain current increases at a lower rate.

3.2.1.6 Effect of varying the drain voltage

Let us now observe the behaviour of the TFET when the drain voltage is varied, for a fixed gate voltage $V_{GS} > 0$. As long as the drain voltage is lower than the gate voltage (i.e. $V_{DS} < V_{GS}$), the channel potential is pinned to the drain potential, as shown in Figure 3.7. As we can observe from these band diagrams, when $0 \leq V_{DS} < V_{GS}$, there is no depletion region at the channel–drain junction. Therefore, there is negligible resistance between the channel and drain, leading to the voltage being nearly constant in this region. The channel potential is thus solely controlled by the drain potential. As the drain bias is increased, the energy bands in the drain as well as in the channel get lowered with respect to the source. This causes an increase in the tunnelling of electrons from the source to the channel, and therefore a significant increase in the drain current.

However, when the drain voltage is increased such that $V_{DS} \sim V_{GS}$, a depletion region starts to form at the channel–drain junction. When this depletion region is of the same length as in the case of thermal equilibrium, the drain control over the channel potential is significantly reduced, as shown in Figure 3.7. Therefore, the device current remains nearly constant with a further increase in the drain bias.

We will now compare the qualitative predictions made in this section with the observed characteristics of a TFET.

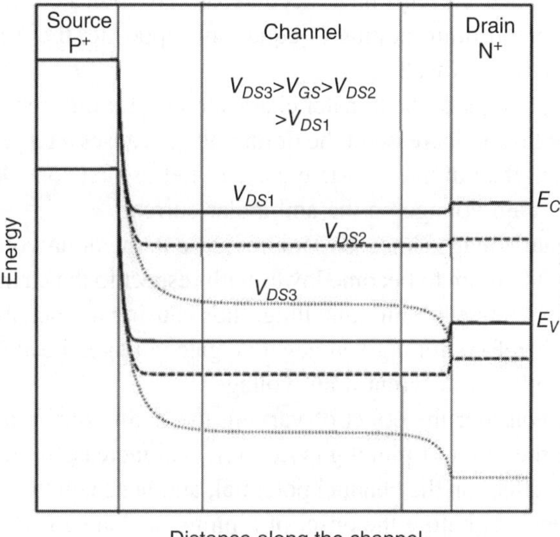

Figure 3.7 Band diagram along the surface of a TFET for a fixed value of the gate voltage (V_{GS}) (in the ON-state) and different values of drain voltage (V_{DS}).

3.2.2 Device characteristics

3.2.2.1 Transfer characteristics

Figure 3.8(a) shows the transfer characteristics of a TFET in which the source and drain dopings are equal and opposite. The regions of operation discussed in the previous section have been marked on the figure. As we can observe, there is a very low current in the OFF-state of the device, when the conduction band of the channel is not aligned with the valence band of the source. As the gate bias is increased, the current rapidly increases due to reduction of the tunnelling width and an increase in the number of initial states in the source from where tunnelling can occur. At a higher gate bias, the rate of increase of current reduces since the contribution of the additional initial states in the source valence band is negligible. As the gate bias approaches the drain potential, the rate of increase of current further decreases due to pinning.

A similar behaviour is observed in the negative bias region. As mentioned in Section 3.2.1.5, this symmetrical behaviour of the TFET is referred to as ambipolar operation and the current curve in the negative bias region is referred to as the ambipolar current of the TFET. Let us briefly explore this ambipolar behaviour of the TFET. The current characteristics under positive and negative gate bias are due to source–channel and drain–channel tunnelling, respectively. The transfer characteristics due to source–channel tunnelling (ON-state current) and drain–channel tunnelling (ambipolar current) have been separately indicated in Figure 3.8(b). Since the source and drain doping is equal and opposite, these two curves are perfectly symmetrical as well.

Figure 3.8(c) compares the transfer characteristics for different drain voltages. We can observe that an increase in the drain voltage causes a larger change in the ambipolar current than in the ON-state current. Let us first consider the effect of the varying the drain voltage on the ambipolar current.

When the gate voltage is negative, an increase in the drain voltage causes the energy bands in the drain to become lower with respect to the channel. This leads to an increase in channel–drain tunnelling, thus causing an increase in the ambipolar current. Therefore, for a given negative gate voltage, the ambipolar current differs significantly for different drain voltages.

Let us now consider the effect of varying the drain voltage on the ON-state current. Before the onset of pinning ($V_{GS} \leq V_{DS}$), an increase in the drain voltage has a negligible effect on the channel potential, and hence on the source–channel tunnelling. Therefore, before the onset of pinning, a change in the drain voltage has a negligible effect on the ON-state current for a given gate voltage. Pinning occurs when $V_{GS} \geq V_{DS}$, in which case the channel potential is fixed at the drain potential. After pinning has occurred, an increase in the drain voltage lowers the channel energy bands with respect to the source. This leads to an increase

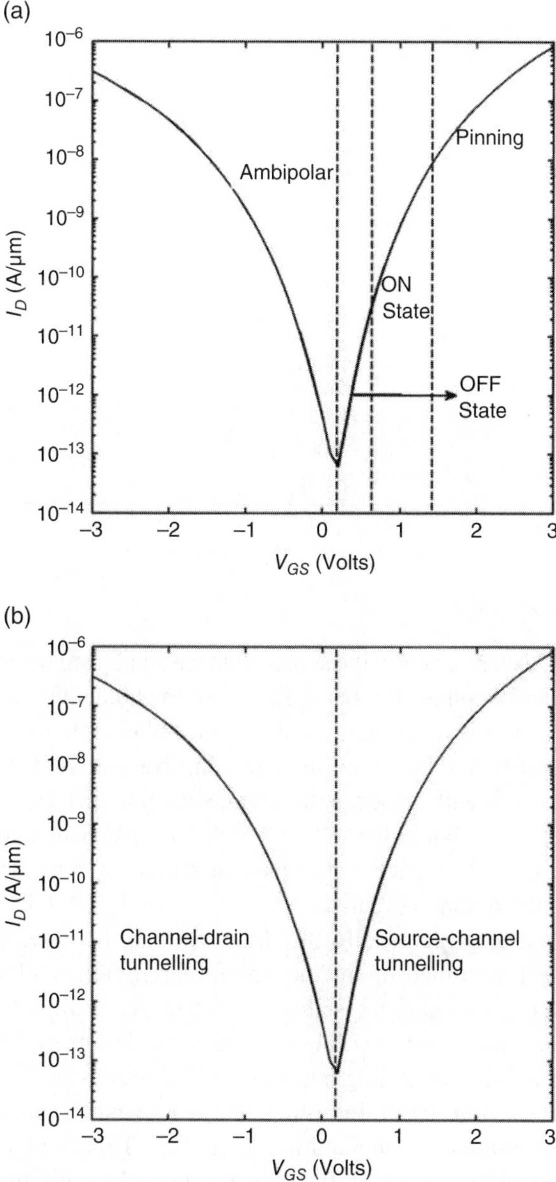

Figure 3.8 Transfer characteristics (I_D –V_{GS}) of an n-channel TFET (a) for a positive value of V_{DS}, showing different regions of operations, (b) showing the source–channel tunnelling (ON-state current) and drain–channel tunnelling (ambipolar current) and (c) for different values of V_{DS}.

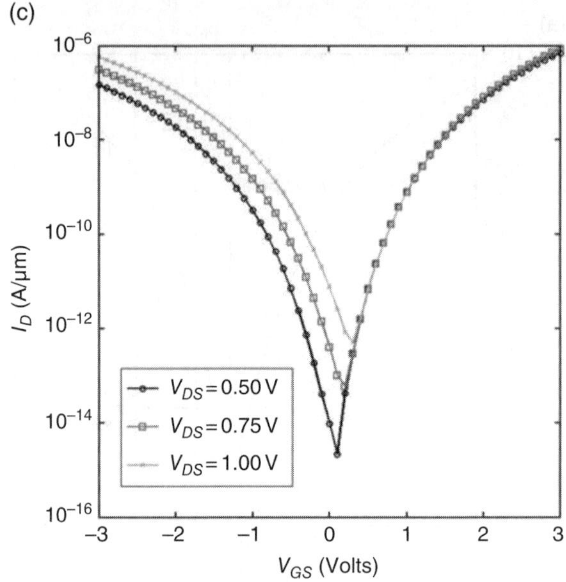

Figure 3.8 (Continued)

in the number of energy states in the source valence band from where electrons can tunnel to the channel conduction band. However, the tunnelling length associated with these energy states is very large and their contribution to the source–channel tunnelling is negligible. Therefore, after pinning has occurred, lowering of the channel energy bands with respect to the source increases the source–channel tunnelling solely due to a shorter tunnelling length. Thus, a change in the drain voltage leads to a smaller difference in the ON-state current after the onset of pinning, as compared to the ambipolar current.

In other words, an increase in the drain voltage increases the number of states available for tunnelling in the ambipolar region, and the tunnelling length for the electrons in these states is smaller. After pinning has occurred in the ON-state, a higher drain voltage increases the number of available states for tunnelling, and the tunnelling length for the electrons in these states is larger. Therefore, varying the drain voltage causes a smaller change in the ON-state current as compared to the ambipolar current. This results in the minima of transfer characteristics to shift towards the right with increasing drain voltage.

3.2.2.2 Subthreshold characteristics of MOSFET versus TFET

The main premise of studying the TFETs as an alternative to the MOSFETs is due to the improved subthreshold swing (SS) exhibited by the TFETs. While

MOSFETs have a thermal limit of 60 mV/dec on SS, the TFETs can achieve an SS below 60 mV/dec as they operate on a different mechanism (i.e. tunnelling) as compared to the MOSFET (i.e. thermionic injection). However, with this difference in the mechanism of current conduction in the subthreshold region comes a difference in the definition of the subthreshold swing.

In the subthreshold region, a MOSFET conducts current by the mechanism of diffusion of carriers across the channel, and can be written as [8]

$$I_{DS}=\mu_{eff}\frac{W}{L}\sqrt{\frac{\varepsilon_{Si}qN_A}{4\psi_B}}\left(\frac{kT}{q}\right)^2 e^{q(V_{GS}-V_{\text{th}})/mkT}\left(1-e^{-qV_{DS}/kT}\right) \tag{3.2}$$

where ψ_B is the Fermi potential of the silicon body and the other terms have their usual meaning.

The subthreshold swing (SS) for a MOSFET is defined as

$$SS=d\log(I_{DS})/dV_{GS} \tag{3.3}$$

Hence, the SS of a MOSFET is constant with V_{GS}.

On the other hand, a TFET, even in the subthreshold region, conducts by the mechanism of band-to-band tunnelling. As seen in Equation (2.53), the tunnelling generation rate is an inverse exponential function of the electric field. Therefore, in a TFET:

$$I_{DS}\propto e^{-1/E} \tag{3.4}$$

where E is the lateral electric field in the channel, which in turn is a function of V_{GS}. Hence, in a TFET, the SS varies with the gate voltage and does not have a constant value as in the case of a MOSFET. Hence, for a TFET, two kinds of SS are defined in general: (i) point subthreshold swing and (ii) average subthreshold swing.

The point subthreshold swing is defined at a particular value of V_{GS}, as given below:

$$SS_{point}(V_{GS})=d\log(I_{DS}(V_{GS}))/dV_{GS} \tag{3.5}$$

The average subthreshold swing is defined over a range of V_{GS} values, as given below:

$$SS_{AVG}=\frac{V_{th}-V_{off}}{\log\left(I_{DS}(V_{th})-I_{DS}\left(V_{off}\right)\right)} \tag{3.6}$$

where V_{th} is the threshold voltage of the TFET and V_{off} is the gate voltage in the OFF-state.

At any point in this text, unless mentioned otherwise, the subthreshold swing in the context of a TFET refers to the SS_{AVG}.

3.2.2.3 Output characteristics

Figure 3.9 shows the output characteristics of a TFET. As discussed in Section 3.2.1.6, initially the gate potential is greater than the drain potential, due to which the channel potential is pinned at the drain potential. Therefore, as the drain potential is increased, the channel potential increases, leading to a significant increase in the current. As the drain potential approaches the gate potential, the channel potential is no longer dependent on the drain potential. Therefore, the current remains nearly constant with increasing drain bias. In this case, the output characteristics are saturated. In the following section, we will compare the output characteristics of a MOSFET and a TFET.

A comparison of the output characteristics of the MOSFET (Figure 3.10) and the TFET (Figure 3.9) indicates two prominent differences. First, the output resistance in the saturation region is much higher in the case of the TFET, since channel length modulation has a negligible effect on the drain current of a TFET. Moreover, in the saturation region ($V_{DS} > V_{GS}$), changing the drain voltage has a negligible effect on the channel potential, due to which the drain current due to source–channel tunnelling remains constant.

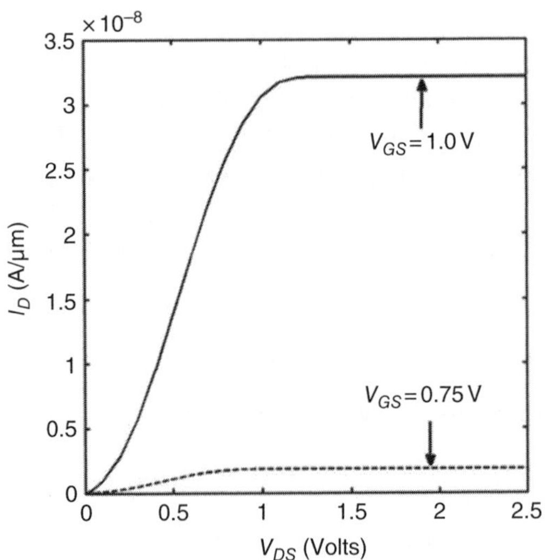

Figure 3.9 Output characteristics (I_D–V_{DS}) of an n-channel TFET for different values of V_{GS}.

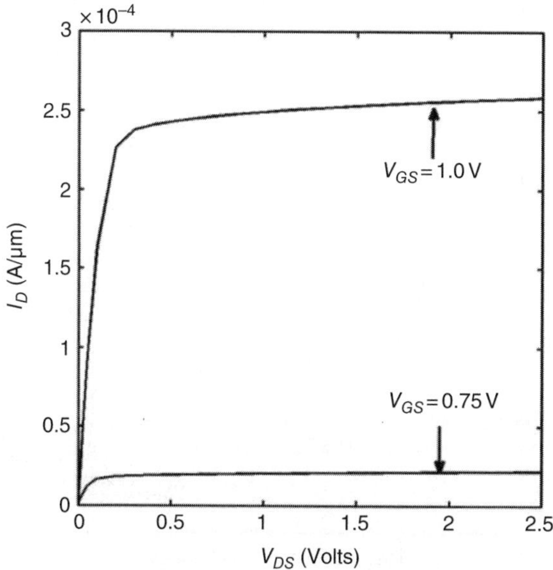

Figure 3.10 Output characteristics (I_D–V_{DS}) of an n-channel MOSFET for different values of V_{GS}.

The other difference in the output characteristics is in the saturation voltage, which is higher in the case of the TFET. This behaviour is referred to as delayed saturation, and makes TFETs less suitable for low-voltage analog applications. Let us examine the physics behind this phenomenon.

3.2.2.4 Delayed saturation in output characteristics

In the output characteristics of a MOSFET, as V_{DS} increases, the drain current saturates. This saturation occurs because, as the drain bias is increased, the electric field at the drain–channel junction becomes large enough to make it reverse-biased, thus causing a pinch-off in the channel at the drain end. Saturation of the drain current with an increase in V_{DS} is also observed in the output characteristics of a TFET. However, in a TFET, the drain voltage at which the drain current saturates (i.e. $V_{DS(\mathrm{sat})}$) is higher as compared to that of a MOSFET with a similar structure, as can be seen in Figure 3.10. Hence, we get a delayed saturation in output characteristics of a TFET. Let us now understand why a TFET exhibits such behaviour in its output characteristics.

Let us have a look at Figure 3.11, which compares the surface potential of the TFET and the MOSFET in the linear region for $V_{GS} > 0$. Surface potential is

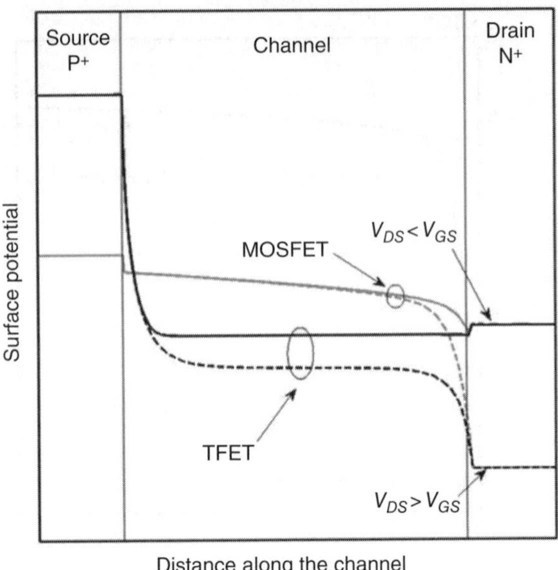

Figure 3.11 Surface potential of a p-channel TFET and a p-channel MOSFET in the linear and the saturation regions of operation.

defined as the potential along the channel of the device at the Si–SiO_2 interface. The potential is measured with respect to the source Fermi level, that is the source Fermi level is taken to be $\psi = 0$. Here we can see that in a TFET, as the drain bias is increased from $V_{DS} = 0$ to $V_{DS} < V_{GS}$, a large part of the drain bias drops across the source–channel junction and only a small part drops across the rest of the channel, whereas in a MOSFET, the drain bias drops across the entire channel almost linearly. Thus the lateral electric field at the drain–channel junction is less in a TFET as compared to that in a MOSFET for the same biasing conditions. However, the magnitude of lateral electric field required at the drain–channel junction to cause the pinch-off of the channel is the same for both TFETs and MOSFETs, since the inversion charge density is identical in the channel. Hence, a greater drain bias (V_{DS}) is needed to cause the channel to be pinched-off in a TFET, leading to an extended linear region in the output characteristics. Thus, the saturation in the output characteristics of a TFET is achieved at a higher V_{DS} as compared to that in a MOSFET with a similar structure.

This behaviour of a TFET can be detrimental in analog applications, as the transistor in an analog circuit works primarily in the saturation region.

A delayed saturation can severely hamper the transconductance of the device and will affect the gain of an amplifier designed using the TFETs. It will also affect the maximum possible output swing of an amplifier (without distorting the output signal). In digital circuits, delayed saturation will affect the switching speed of the logic gate designed using the TFET, as the transistor will not be operating at the maximum possible drain current for a given supply voltage if the saturation voltage $V_{DS(Sat)}$ is greater than the supply voltage V_{DD}. Hence, while designing a TFET for CMOS application, one should make sure that the saturation in the output characteristics of the TFET takes place at $V_{DS} < V_{GS}$, so that at $V_{DS} = V_{DD}$, the device is operating at the maximum possible current.

The characteristics that we have studied in this section are dependent on the particular parameters chosen while fabricating the TFET, such as the dopings, the gate work function, etc. Variation of these parameters allows us to modify the TFET according to the final specifications that are desired. In the next section, we will briefly examine these dependencies.

3.2.3 Performance dependence on device parameters

3.2.3.1 Doping

As already mentioned while comparing the structure of a TFET with that of a MOSFET, the primary difference between the two is the source doping. The source doping in an n-channel TFET is p^+-type. The reason for this can be understood by comparing the band diagrams of the two TFET structures – one with a high source doping N_{A1} and the other with a lower source doping N_{A2}, as shown in Figure 3.12. As we can observe, a higher doping in the source leads to a shorter depletion region at the source–channel junction. Due to this, as indicated in Figure 3.12, the electron occupying the valence band edge in the source would have to tunnel through a shorter distance L_{t_1} in the case of higher source doping, as compared to the larger distance L_{t_2} in the case of lower source doping. Therefore, the tunnelling probability would be greater in the case of a higher source doping, thereby resulting in a larger ON-state current.

Following a similar reasoning, an increase in the drain side doping would lead to a greater drain–channel tunnelling in the case of negative gate bias, thereby increasing the ambipolar current. In many applications, the ambipolar current is sought to be minimised, which may be accomplished by reducing the drain doping. The transfer characteristics of two TFETs is compared in Figure 3.13 – one with a high drain doping and one with a lower drain doping. We find that the ambipolar current is successfully reduced by reducing the drain doping.

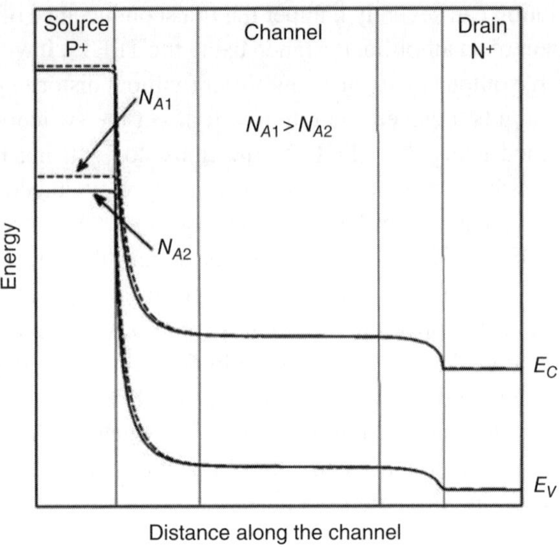

Figure 3.12 Band diagrams along the surface of an n-channel TFET with different values of source doping concentration.

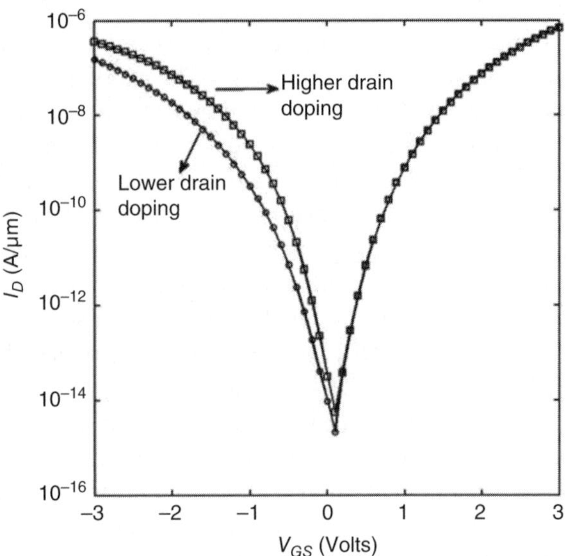

Figure 3.13 Transfer characteristics of an n-channel TFET with different drain doping concentrations for a fixed drain voltage (V_{DS}).

3.2.3.2 Gate work function

In Figure 3.14, we have compared the band diagram of a TFET with a higher gate work function Φ_{G1} to that of a TFET with a lower gate work function Φ_{G2}, at thermal equilibrium. Due to the formation of a greater inversion charge layer in the latter case, the energy bands of the channel are lower. This leads to a steeper source–channel junction, thereby (as discussed in the preceding section) increasing the source–channel tunnelling and the ON-state current. It should be noted that this increase in the ON-state current is accompanied by an increase in the OFF-state current, and thus an increased power dissipation in the OFF-state. Moreover, the drain–channel tunnelling is reduced in this case, since the channel energy bands are now closer to the drain energy bands (Figure 3.14). This behaviour can be observed by comparing the transfer characteristics of these two TFET structures in Figure 3.15.

Another effect of changing the gate work function on the transfer characteristics can be observed from the transfer characteristics (Figure 3.15) – shifting of the central point of the current curve to the left upon a decrease in the gate work function. As discussed in Section 3.2.2.1 of this chapter, the transfer characteristics of the TFET can be viewed as a superposition of two curves – the left-hand side curve due to the drain–channel tunnelling and the right-hand curve due to the source–channel tunnelling. Decreasing the gate work function increases the

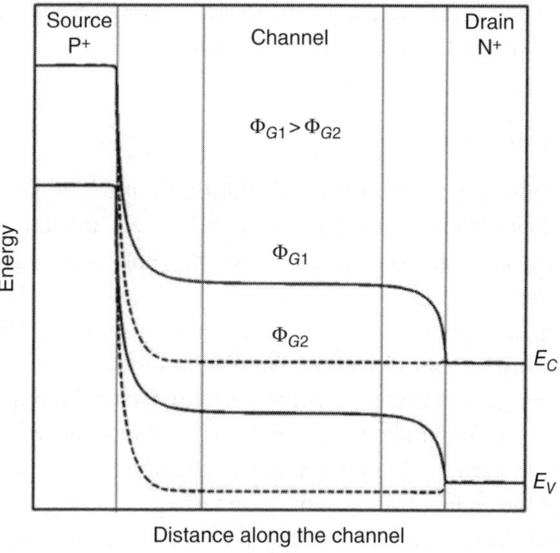

Figure 3.14 Band diagram along the surface of an n-channel TFET for different values of the gate metal work function.

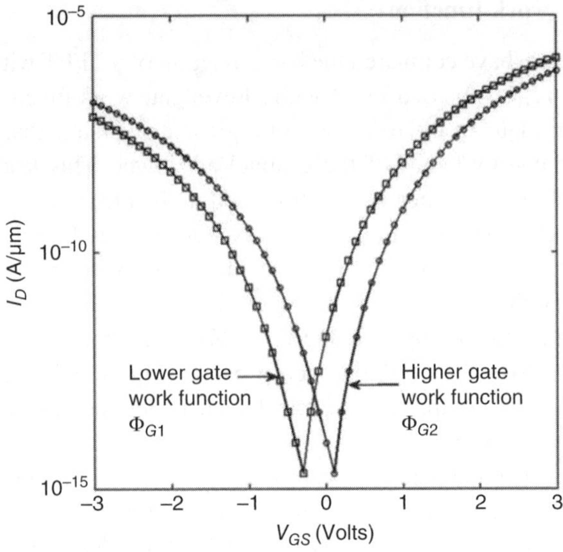

Figure 3.15 Transfer characteristics of an n-channel TFET with different values of the gate work function.

source–channel tunnelling and decreases the drain–channel tunnelling, raising the right-hand curve and lowering the left-hand curve along the current axis, thereby shifting the central point of the curve to the left.

3.2.3.3 Gate oxide

The function of the gate oxide (in both MOSFETs and TFETs) is twofold – to insulate the gate and to enable the gate to control the electrostatics in the channel. Since the oxide is an insulator, the gate oxide prevents current leakage from the channel to the gate.

The latter function of the gate oxide is enhanced by choosing thin oxides with a high dielectric constant. Let us examine this point further. The oxide capacitance is given as $C_{ox} = \epsilon_{ox}/t_{ox}$, where ϵ_{ox} is the permittivity of the oxide and t_{ox} is the thickness. As this formula shows, the gate oxide capacitance can be increased by reducing the thickness. A greater oxide capacitance leads to a higher charge being formed in the channel at the same gate voltage, leading to a greater control of the gate on the electrostatics of the channel. However, a thinner gate oxide may lead to some carriers from the channel tunnelling to the gate, generating a gate leakage current [9], which is not desirable. This can be prevented by the use of a material with a high permittivity, such as an oxide, in addition to a thin layer

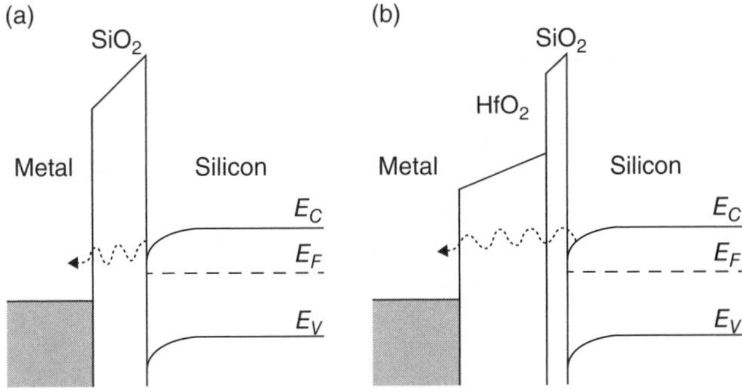

Figure 3.16 Gate leakage in a TFET through (a) SiO_2 and (b) HfO_2/SiO_2 gate stack [9]. Source: Reproduced with permission of APEX/JJAP, Chaturvedi and M. J. Kumar, "Impact of gate leakage considerations in tunnel field effect transistor design", Japanese Journal of Applied Physics, vol. 53: 7, pp.074201-1-8, June 2014.

of silicon dioxide. As shown in Figure 3.16, this would cause a thicker potential barrier between the channel and the gate, thereby reducing the tunnelling rate. The use of an oxide material with high permittivity has the added benefit of further reducing the oxide capacitance and thus increasing the gate control of the channel.

While varying the structural parameters of a TFET results in a significant modification of TFET characteristics, the use of different structures can further enhance the device behaviour. We will now briefly look at commonly studied TFET structures and the advantages that these structures provide when compared to a basic TFET.

3.3 Types of TFETs

Based on the structure, TFETs can be broadly classified into two categories: planar and three-dimensional structures. A planar TFET is a device in which the current-carrying surface is planar. The device can be made on a bulk silicon wafer or on an SOI wafer. For better gate control over the channel, SOI TFETs are preferred over bulk TFETs, and only the former are extensively studied. Let us first study certain important planar TFET structures.

3.3.1 Planar TFETs

Figure 3.17 shows the structure of an SOI TFET. It consists of a thin layer of silicon (typically around 10 nm or less) grown on a layer of buried oxide (around a hundred nm thick) on a silicon substrate. The gate oxide (1–2 nm thick) is grown

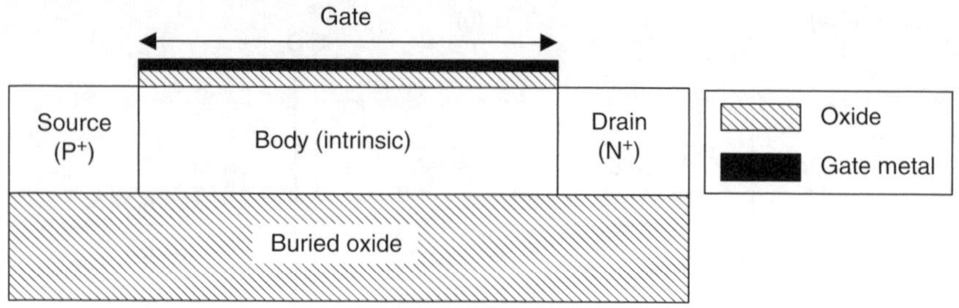

Figure 3.17 Schematic of a p-channel SOI TFET.

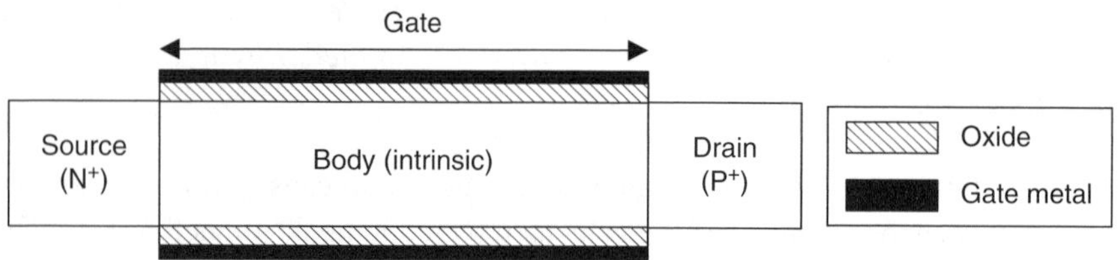

Figure 3.18 Schematic of a p-channel double gate (DG) TFET.

on the thin silicon layer followed by deposition of gate, source and drain metal contacts. The source and drain are then formed by the appropriate doping in the thin silicon layer. In an SOI TFET, the entire thin silicon layer is depleted and the buried oxide layer blocks all the source-to-drain leakage paths through the bulk region. Moreover, the drain-body and source-body depletion regions are small, thus providing a better gate control over the channel. This also leads to a smaller source/drain to substrate capacitance. The advantages of an SOI structure can be further enhanced using the following device structures.

3.3.1.1 Double gate TFET

Figure 3.18 shows the structure of a double gate TFET. It consists of two gates, one at the top (called the front gate) and the other at the bottom (called the back gate). This configuration improves the electrostatic control of the gate on the channel since now the field lines from the gate terminate at the back gate rather than terminating in the channel. The ON-state current is also increased as compared to a single gate TFET, since there are two channels in which current can flow in the device.

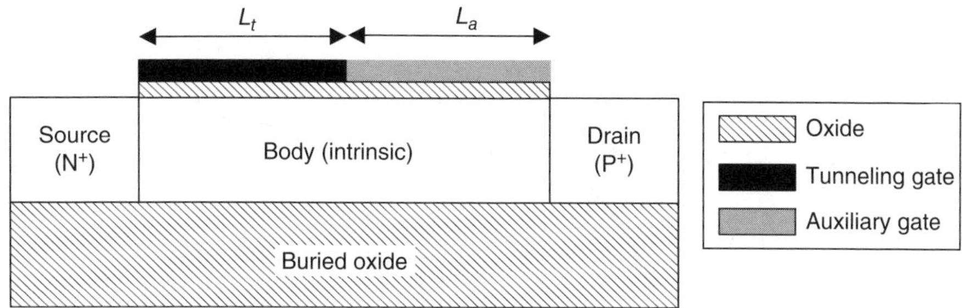

Figure 3.19 Schematic of a p-channel dual material gate (DMG) TFET [10]. Source: Reproduced with permission of IEEE.

3.3.1.2 Dual material gate TFET

An important variation of the TFET structure is the dual material gate (DMG) TFET [10]. A schematic of this device is shown in Figure 3.19. It consists of two gates of different work functions along the length of the channel – one gate covers the part of the channel near the source and the second gate the part near the drain. The gate near the source is called the tunnelling gate, as tunnelling in the ON-state of the device occurs at the source–channel junction. The gate near the drain is called the auxiliary gate. The DMG structure can be fabricated using sidewall spacer techniques [11–13].

In a p-type DMG TFET, the tunnelling gate has a higher work function as compared to the auxiliary gate. As discussed in Section 3.2.3.1, this leads to a higher potential difference between the source and the channel, thus reducing the tunnelling length and increasing the ON-state current. If the length of the tunnelling gate is small in comparison with the auxiliary gate, the surface potential in the OFF-state depends solely on the auxiliary gate work function. As the auxiliary gate has a lower work function, the potential difference between the source and the channel is lower in the OFF-state, thus reducing the OFF-state current. This behaviour is displayed in Figure 3.20, which compares the transfer characteristics of two single material gate (SMG) TFETs having different gate work functions with that of a DMG TFET. The SMG with a higher work function (4.8 eV) has a higher ON-state current than the SMG with a lower work function (4.4 eV). However, this higher ON-state current is accompanied by a correspondingly higher OFF-state current, leading to a higher subthreshold swing and a greater power dissipation in the OFF-state. While the SMG with the lower gate work function has a lower subthreshold swing, its ON-state current is low. The DMG is able to combine the benefits of both these structures – it has an ON-state current equal to an SMG with a gate work function of 4.8 eV and an OFF-state current equal to that of the SMG

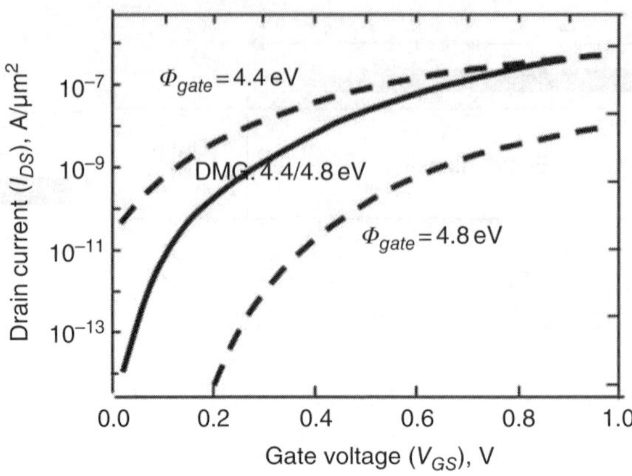

Figure 3.20 Comparison of the transfer characteristics of two single material gate (SMG) TFETs having different gate work functions to that of a DMG TFET.

with a gate work function of 4.4 eV. Therefore, the DMG TFET provides a higher subthreshold slope as compared to a conventional SMG TFET as well as a higher ON-state current. It also has lower DIBL effects as the channel under the auxiliary gate has a higher resistance as compared to the channel under the tunnelling gate, causing a lower potential drop in the latter and shielding it from the variations in the drain bias [10, 14]. It also has less threshold voltage roll-off.

3.3.1.3 p-n-p-n TFET or n-p-n-p TFET

Another popular TFET structure is the p-n-p-n TFET [15–18], the structure of which is shown in Figure 3.21. It consists of a thin P^+ pocket at the source–channel junction. This pocket leads to the formation of a local minima at the conduction band edge *EC* at *VGS* = 0 V, as shown in Figure 3.22, which results in a more abrupt change and a lower value of tunnelling barrier width. It also increases the lateral electric field at the tunnelling junction, giving a higher ON-state current. Apart from this, a reduction in the vertical electric field minimises the generation rate of the interface traps, improving device reliability. For obtaining a steep subthreshold slope, an optimum length (~4 nm) of the pocket-doped region is needed. If the pocket is too long, it is not fully depleted, and the tunnelling junction is not gate controlled. This degrades the subthreshold swing of the device. Figure 3.23 shows the band diagram of p-n-p-n TFET with varying widths of the pocket. Fabricating a narrow doped pocket and ensuring the steepness of the junction is difficult. It is possible either with an epitaxial growth in a vertical transistor or a very precise implant. One possible solution to this difficulty is to use the charge plasma

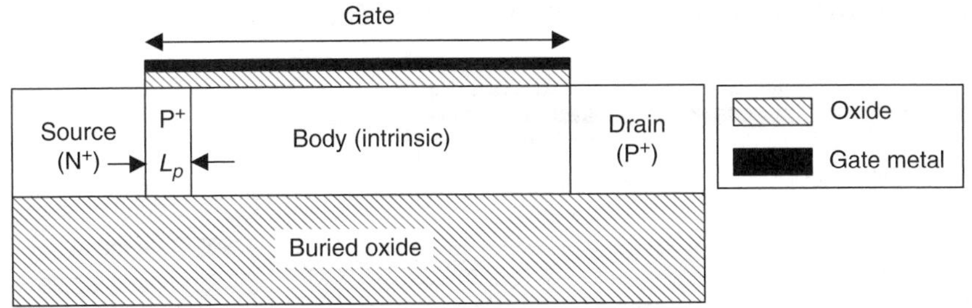

Figure 3.21 Schematic of a p-channel n-p-n-p TFET (i.e. p-n-p-n TFET for an n-channel TFET).

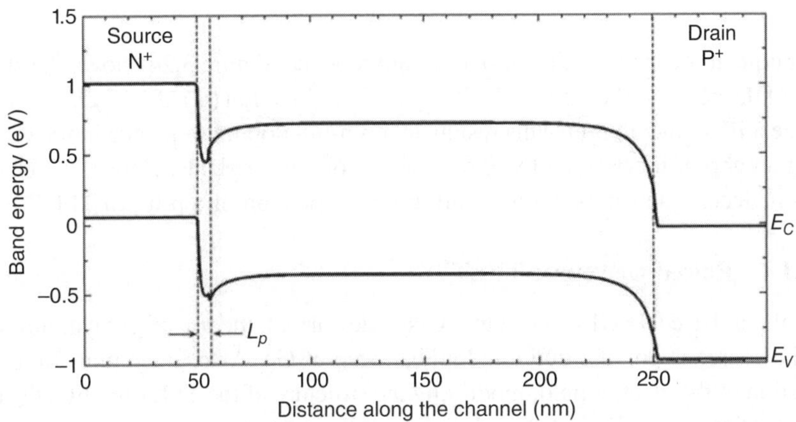

Figure 3.22 Band diagram of a p-channel n-p-n-p TFET (i.e. p-n-p-n TFET for an n-channel TFET) along the surface in the ON-state.

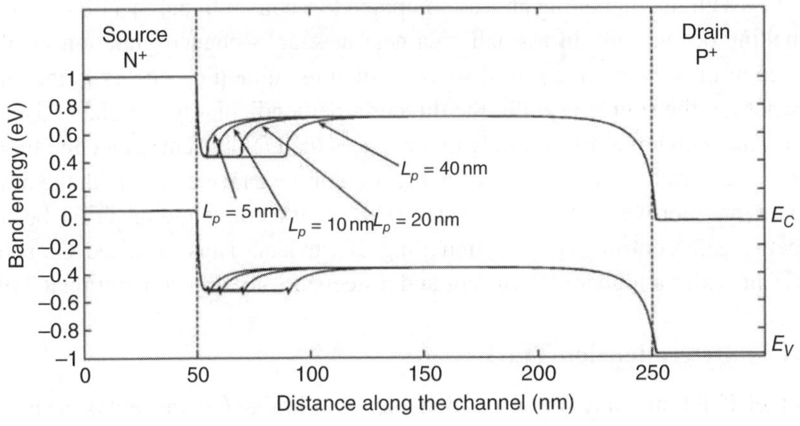

Figure 3.23 Band diagram of a p-channel n-p-n-p TFET (i.e. p-n-p-n TFET for an n-channel TFET) along the surface with varying widths of the P+ pocket region.

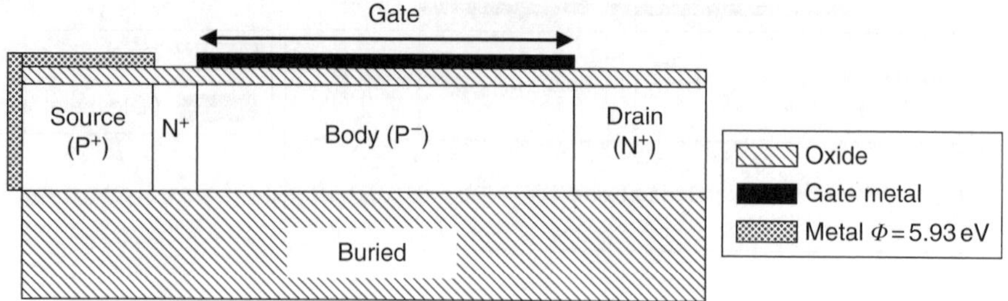

Figure 3.24 Schematic of a charge plasma p-n-p-n TFET.

technique to generate the p-n-p-n doping sequence. Figure 3.24 shows the structure of the device, where a metal of the gate work function (Φ) of 5.93 eV is used to create a P^+ source region. This results in a p-n-p-n doping sequence in the device with an abrupt junction and without the use of any implants. Hence, the charge plasma technique can be very useful in the fabrication of a p-n-p-n TFET.

3.3.1.4 Raised Ge-source TFET

As silicon-based TFETs have a low ON-state current, the use of germanium (Ge) in the source is popular due to a low bandgap of Ge. Also, since tunnelling is a function of the electric field, modifying the structure of the TFET to align the direction of the tunnelling of the carriers with the gate electric field is also a good option [19]. Hence, a raised germanium source TFET as shown in Figure 3.25 is an important TFET structure and provides many advantages over a conventional Si-based TFET structure. As can be seen in Figure 3.25, the raised source structure provides a higher tunnelling area as compared to a conventional Si TFET in which tunnelling occurs only in a small area near at source–channel junction near the surface of the device. In a raised source structure, tunnelling occurs in the entire thickness of the source (T_{Ge}) in the direction perpendicular to the gate. Also, the use of Ge, which has a low bandgap, increases the ON-current. Since the tunnelling of the carriers is aligned with the direction of the gate electric field, the device also has an improved subthreshold slope (SS) over a conventional TFET because of better gate control over the tunnelling of carriers. Thus, a raised Ge-source TFET provides a higher ON-current and a steeper SS over a conventional TFET.

3.3.1.5 Heterojunction TFET

Another TFET structure proposed for achieving a higher ON-current is the heterojunction TFET [20], made up of III–V materials. Figure 3.26(a) shows the structure of a heterojunction TFET. It consists of $GaAs_{0.35}Sb_{0.65}$ as the source material

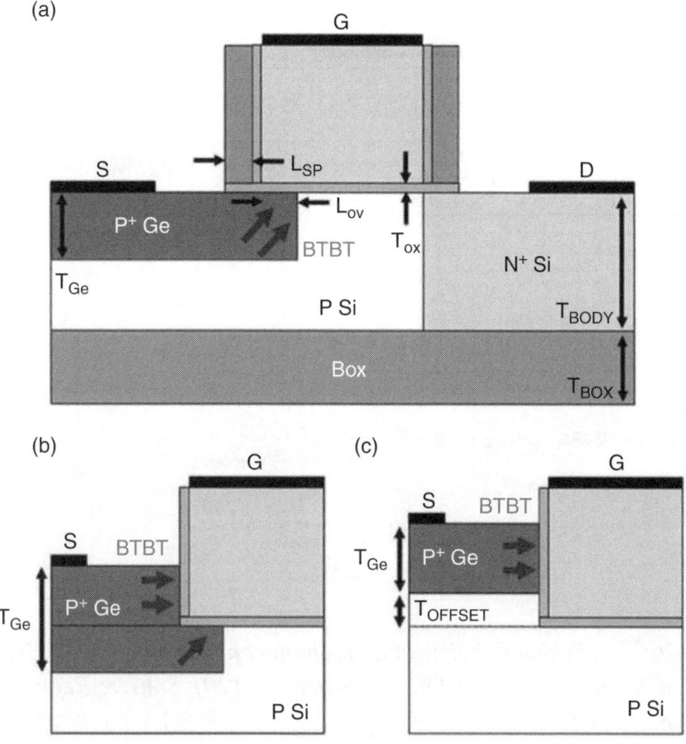

Figure 3.25 Schematic of (a) planar, (b) partially raised and (c) fully raised Ge-source TFETs. Dominant directions of tunnelling are shown by arrows [19].

and $In_{0.7}Ga_{0.3}As$ as the channel and drain material, giving us a heteromaterial junction at the source– channel interface. Figure 3.26(b) shows the band diagram of such a device at the surface. As can be seen in Figure 3.26(b), the band diagram of a heterojunction is staggered at the source–channel junction. This leads to a shorter tunnelling width, thus leading to an increased ON-current. Also, III–V materials are direct bandgap materials and have a bandgap smaller than Si. This also contributes to the increased ON-state current in heterojunction TFETs.

3.3.1.6 Ferroelectric TFET

A structure proposed for improving the SS and the ON-state current of a TFET is the ferroelectric TFET. In this structure a ferroelectric material in used in the gate stack as shown in Figure 3.27. When a gate voltage is applied, the polarisation of the ferroelectric material in the gate stack increases the effective gate voltage seen by the channel [21]. In other words, the vertical electric field across the gate oxide increases, thus causing more inversion on the channel for a particular value of

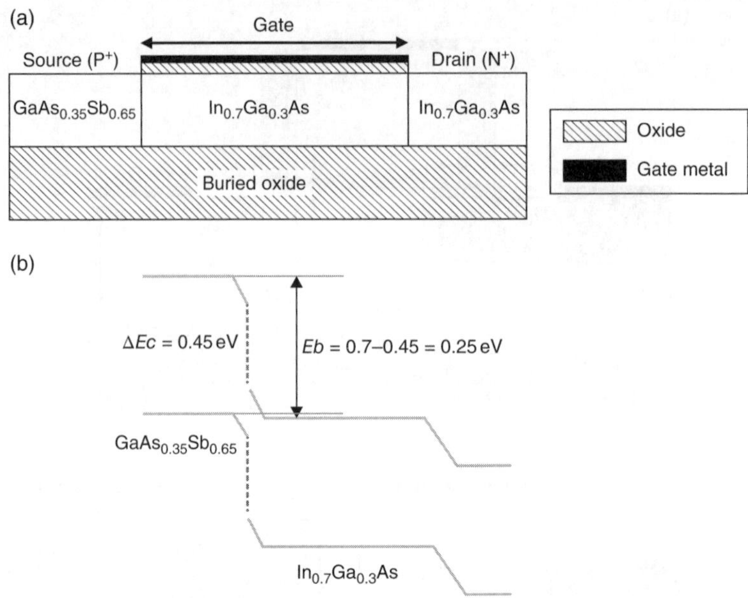

Figure 3.26 (a) Schematic view of an n-channel heterojunction TFET. (b) Band diagram of the heterojunction TFET shown in (a) [20]. Source: Reproduced with permission of IEEE.

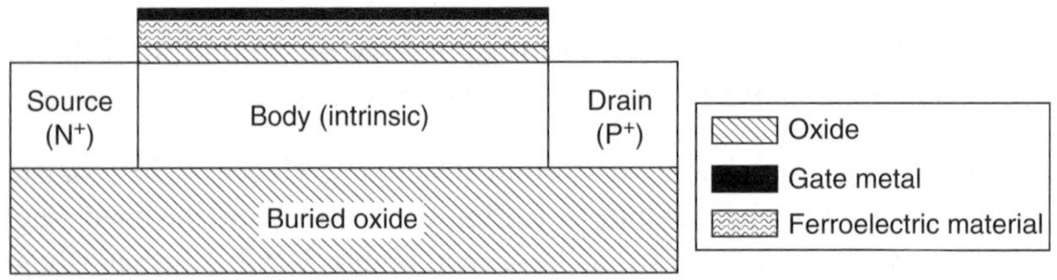

Figure 3.27 Schematic view of a p-channel ferroelectric TFET.

applied gate voltage. Thus, the steepness of the I_{OFF} to I_{ON} transition increases, improving the SS and the ON-current. Figure 3.28 gives a qualitative description of the transfer characteristics of a ferroelectric TFET and compares them with that of a MOSFET and a conventional TFET.

3.3.2 Three-dimensional TFETs

A three-dimensional TFET is a device in which the current-carrying surface extends in all three dimensions. Two of the most important three-dimensional structures are the gate all around nanowire TFET and the tri-gate TFET.

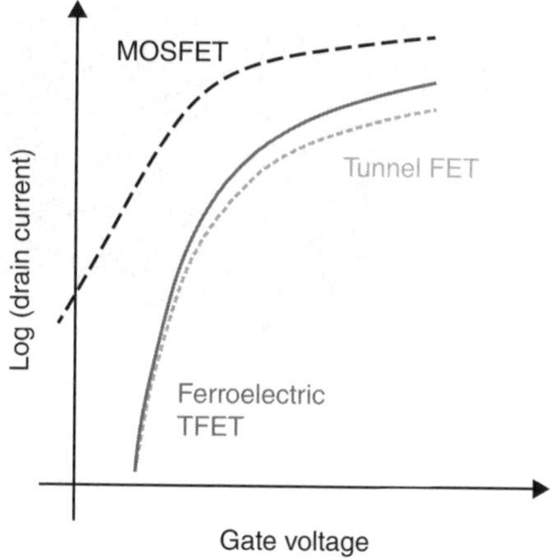

Figure 3.28 Comparison of the I_D–V_{GS} curves of a ferroelectric TFET with that of a MOSFET and a conventional TFET [21].

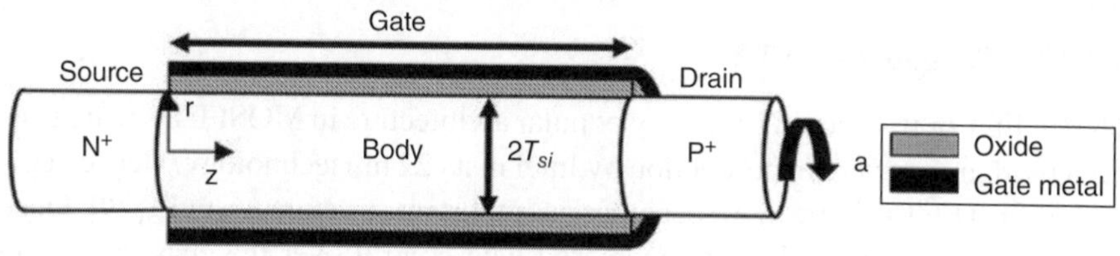

Figure 3.29 Schematic view of a gate all around (GAA) nanowire TFET.

3.3.2.1 Gate all around nanowire TFET

The increased gate control observed in a double gate TFET is further improved in a gate all around (GAA) TFET [22–28]. Figure 3.29 shows the structure of a gate all around nanowire TFET. It consists of a silicon nanowire of radius 10 nm, surrounded by a gate oxide of thickness 2 nm, covered all around by the gate metal. Such a structure can achieve a high level of electrostatic control of the gate, as all the field lines originating from the drain can terminate at the gate, without significant penetration into the channel. This leads to a steeper subthreshold slope and diminished short channel effects such as DIBL and threshold voltage roll-off. It also enhances the ON-state current, as this geometry provides a large area for the current to flow as compared to a planar device.

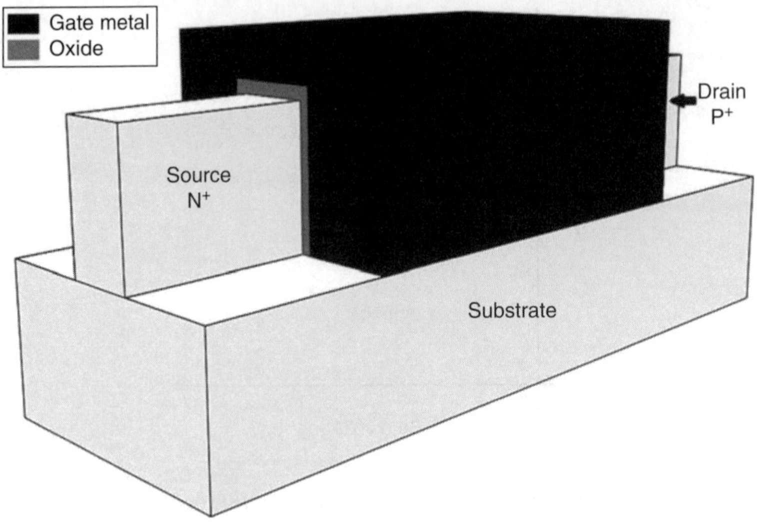

Figure 3.30 Schematic view of a tri-gate (fin) TFET.

3.3.2.2 Tri-gate/fin TFET

The fin FET or tri-gate FET is a very popular architecture in MOSFETs, which has been used in commercial production by Intel in its 22 nm technology. Hence, a tri-gate or fin TFET (Figure 3.30) becomes an important structure to study [29]. Once again, this structure leads to an increased gate control over the channel and an increased area of current flow. This leads to a higher ON-state current, a steeper subthreshold slope and diminished short channel effects. These improvements provided by the tri-gate TFET are lesser than those provided by the GAA nanowire TFET, but fabrication of a tri-gate TFET is comparatively simpler. Moreover, as tri-gate MOSFETs are already used in commercial production, there is a greater chance for commercial production of tri-gate TFETs.

3.3.3 Carbon nanotube and graphene TFETs

The primary motivation behind the design of TFETs is to fabricate a device that exhibits a sub 60 mV/decade subthreshold swing over a wide range of V_{GS}, and having an ON-state current that is of the order of 1–10 μA/μm. It is difficult to realise silicon TFETs that consistently exhibit an ON-current of the order of silicon MOSFETs used in CMOS technology, while also having a subthreshold swing of sub-60 mV/decade. While ideally simulated devices exhibit a low subthreshold swing (SS) for high ON-state currents, the experimental demonstration of the same has been limited. Si and Ge TFETs are not able to achieve a high

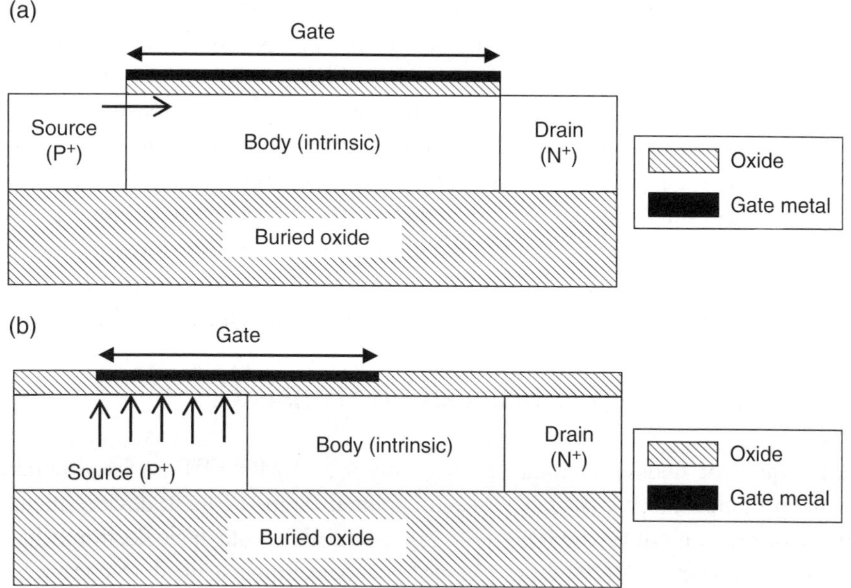

Figure 3.31 (a) Point tunnelling and (b) line tunnelling in a TFET.

ON-state current due to their indirect bandgap. III–V semiconductors are predicted to have high ON-state currents since they have direct bandgaps, yet the experimental realisations of such TFETs do not yet show low SS for a large range of V_{GS}. Two-dimensional materials are the most promising candidates for the fabrication of such devices. A lack of dangling surface bonds in such materials contributes to a higher degree of gate control, and their high mobilities and low bandgaps, especially in the case of graphene, are promising for both high ON-state and low SS applications. Due to a combination of these factors, carbon nanotube (CNT) TFETs and graphene nanoribbon TFETs are theoretically predicted to have the best performance at low gate lengths. In general, the fabrication of TFETs exhibits a trade-off between high ON-state currents and low SS [30].

3.3.4 Point versus line tunnelling in TFETs

The most general problem exhibited by TFETs is their low ON-state current. The ON-state current of a conventional Si-based TFET is of the order of 10^{-6} A/μm. To boost the ON-state current of a TFET, one of the most common approaches is to increase the effective volume of the tunnelling region. In a conventional TFET structure (shown in Figure 3.31(a), the tunnelling of carriers takes place in a small region at the surface near the source–channel junction, in the direction along the channel. This kind of tunnelling at the source–channel junction is known as point tunnelling. One of the ways to increase this volume of tunnelling region is to overlap the gate over the source region and make the carriers tunnel in the vertical direction towards the gate [31–33] (shown in Figure 3.31(b)). This technique increases the effective volume of the tunnelling region, thus increasing the drain current. The carriers due to the high vertical electrical field of the gate tunnel within the source region are then swept away towards the drain due to the lateral electric field of the drain bias. This structural configuration of tunnelling is known as line tunnelling. One example of a line tunnelling FET that you are already familiar with is the raised-Ge source TFET, discussed in Section 3.3.1.4. All the other TFETs discussed in Section 3.3 are point tunnelling TFETs.

3.4 Other steep subthreshold transistors

In the quest of finding alternatives for highly scaled MOSFETs, a large number of steep subthreshold devices have been explored. Some of these devices include impact-ionisation MOS (I-MOS) [34–38], MOSFETs with nanoelectromechanical gate electrodes (NEMFETs) [39–41], nanoelectromechanical (NEM) relays [42, 43], ferroelectric FETs [44, 45] and feedback FETs [46]. All these devices are capable of achieving a subthreshold swing below 60 mV/decade at room temperature. However, most of these devices exhibit a low SS only for a small range of gate voltages and not for the entire subthreshold region. I-MOS, NEMFETs and NEM relays have a steep SS only in a local region and their average SS is much degraded. Moreover, NEMFETs and NEM relays have an ON-state current lower than the conventional TFET. Devices such as I-MOS are not suitable for low-voltage applications since they need a supply voltage more than 1.0 V. Further, feedback FETs and ferroelectric FETs exhibit a degraded output resistance [47]. Also, all of these devices require a complex process of fabrication and, hence, suffer from both yield and reliability issues.

References

[1] A. C. Seabaugh and Q. Zhang, "Low-Voltage Tunnel Transistors for Beyond CMOS Logic", *Proc. IEEE*, vol. 98, no. 12, pp. 2095–2110, December 2010.

[2] A. M. Ionescu and H. Riel, "Tunneling Field-Effect Transistors as Energy-Efficient Electronic Switches", *Nature*, vol. 479, pp. 329–337, November 2011.

[3] S. Saurabh and M. J. Kumar, "Impact of Strain on Drain Current and Threshold Voltage of Nanoscale Double Gate Tunnel Field Effect Transistor (TFET): Theoretical Investigation and Analysis", *Japanese Journal of Applied Physics*, vol. 48, paper no. 064503, June 2009.

[4] S. Saurabh and M. J. Kumar, "Estimation and Compensation of Process Induced Variations in Nanoscale Tunnel Field Effect Transistors (TFETs) for Improved Reliability", *IEEE Trans. on Device and Materials Reliability*, vol. 10, pp. 390–395, September 2010.

[5] M. J. Kumar and S. Janardhanan, "Doping-less Tunnel Field Effect Transistor: Design and Investigation", *IEEE Trans. Electron Devices*, vol. 60, pp. 3285–3290, October 2013.

[6] M. S. Ram and D. B. Abdi, "Single Grain Boundary Tunnel Field Effect Transistors on Recrystallized Polycrystalline Silicon: Proposal and Investigation", *IEEE Electron Device Letters*, vol. 35, no. 10, pp. 989–992, October 2014.

[7] A. Chaudhry and M. J. Kumar, "Controlling Short-Channel Effects in Deep Submicron SOI MOSFETs for Improved Reliability: A Review", *IEEE Trans. on Device and Materials Reliability*, vol. 4, pp. 569–574, April 2004.

[8] Y. Taur and T. H. Ning, *Fundamentals of Modern VLSI Devices*, Cambridge University Press, Cambridge, UK, p. 128, 1998.

[9] P. Chaturvedi and M. J. Kumar, "Impact of Gate Leakage Considerations in Tunnel Field Effect Transistor Design", *Japanese Journal of Applied Physics*, vol. 53, pp. 074201-1–8, June 2014.

[10] S. Saurabh and M. J. Kumar, "Investigation of the Novel Attributes of a Dual Material Gate Nanoscale Tunnel Field Effect Transistor", *IEEE Trans. Electron Devices*, vol. 58, no. 2, pp. 404–410, 2011.

[11] H. Lou, L. Zhang, Y. Zhu, X. Lin, S. Yang, J. He and M. Chan, "A Junctionless Nanowire Transistor with a Dual-Material Gate", *IEEE Trans. Electron Devices*, vol. 59, no. 7, pp. 1829–1836, 2012.

[12] M. J. Lee and W. Y. Choi, "Effects of Device Geometry on Hetero-Gate-Dielectric Tunneling Field-Effect Transistors", *IEEE Electron Device Lett.*, vol. 33, no. 10, pp. 1459–1461, 2012.

[13] M.-L. Fan, V. P. Hu, Y.-N. Chen, P. Su, and C.-T. Chuang, "Analysis of Single-Trap-Induced Random Telegraph Noise and Its Interaction with Work Function Variation for Tunnel FET", *IEEE Trans. Electron Devices*, vol. 60, no. 6, pp. 2038–2044, 2013.

[14] R. Vishnoi and M. J. Kumar, "Compact Analytical Model of Dual Material Gate Tunneling Field Effect Transistor Using Interband Tunneling and Channel Transport", *IEEE Trans. Electron Devices*, vol. 61, no. 6, pp. 1936–1942, 2014.

[15] V. Nagavarapu, R. Jhaveri and J. C. S Woo, "The Tunnel Source (PNPN) n-MOSFET: A Novel High Performance Transistor", *IEEE Trans. on Electron Devices*, vol. 55, no. 4, 2008.

[16] D. B. Abdi and M. J. Kumar, "In-built N+ Pocket PNPN Tunnel-Field Effect Transistor", *IEEE Electron Device Lett.*, vol. 35, no. 12, pp. 1170–1172, 2014.

[17] D. B. Abdi and M. J. Kumar, "PNPN Tunnel FET with Controllable Drain Side Tunnel Barrier Width: Proposal and Analysis", *Superlattices and Microstructures*, vol. 86, pp. 121–125, October 2015.

[18] M. S. Ram and D. B. Abdi, "Single Grain Boundary Dopingless PNPN Tunnel FET on Recrystallized Polysilicon: Proposal and Theoretical Analysis", *IEEE Journal of Electron Devices Society*, vol. 3, no. 3, pp. 291–296, May 2015.

[19] S. H. Kim, S. Agarwal, Z. A. Jacobson, P. Matheu, C. Hu and T-J. K. Liu, "Tunnel Field Effect Transistor with Raised Germanium Source", *IEEE Electron Device Lett.*, vol. 31, no. 10, pp. 1107–1109, 2010.

[20] D. K. Mohata, R. Bijesh, S. Majumdar, C. Eaton, R. Engel-Herbert, T. Mayer, V. Narayanan, J. M. Fastenau, D. Loubychev, A. K. Liu and S. Dutta, "Demonstration of MOSFET-like On-Current Performance in Arsenide/Antimonide Tunnel FETs with Staggered Hetero-junctions for 300 mV Logic Applications", in *IEEE Intl. Electron Dev. Meeting*, pp. 33.5.1–4, December 2011.

[21] A. M. Ionescu, L. Lattanzio, G. A. Salvatore, L. D. Michielis, K. Boucart and D. Bouvet, "The Hysteretic Ferroelectric Tunnel FET", *IEEE Transactions on Electron Devices*, vol. 57, no. 12, pp. 3518–3524, December 2010.

[22] Q. Shao, C. Zhao, C. Wu, J. Zhang, L. Zhang and Z. Yu, "Compact Model and Projection of Silicon Nanowire Tunneling Transistors (NW-tFETs)", in *Int. Conf. of Elec. Dev. and Solid-State Circuits (EDSSC)*, June 2013, pp. 1–2.

[23] A. S. Verhulst, B. Sorée, D. Leonelli, W. G. Vandenberghe and G. Groeseneken, "Modeling the single-gate, double-gate, and gate-all-around tunnel field-effect transistor", *J. Appl. Phys.*, vol. 107, pp. 024518-1-024518-6, January 2010.

[24] H. S. P Wong, "Beyond the Conventional Transistor", *IBM J. of Research and Development*, vol. 46, no. 2/3, pp. 133–168, March/May 2002.

[25] N. Jain, E. Tutuc, S. K. Banerjee and L. F. Register, "Performance Analysis of Germanium Nanowire Tunneling Field Effect Transistors", in *Device Research Conference*, June 2008, pp. 99–100.

[26] A. Zhan, J. Mei, L. Zhang, H. He, J. He and M. Chan, "Numerical Study on Dual Material Gate Nanowire Tunnel Field-Effect Transistor", in *Int. Conf. on Elec. Devices and Solid State Circuit (EDSSC)*, 2012, pp. 1–5.

[27] R. Vishnoi and M. J. Kumar, "A Pseudo 2D-analytical Model of Dual Material Gate All-Around Nanowire Tunneling FET", *IEEE Transactions on Electron Devices*, vol. 61, no. 7, pp. 2264–2270, July 2014.

[28] R. Vishnoi and M. J. Kumar, "Compact Analytical Drain Current Model of Gate-All-Around Nanowire Tunneling FET", *IEEE Transactions on Electron Devices*, vol. 61, no. 7, pp. 2599–2603, July 2014.

[29] D. Leonelli, A. Vandooren, R. Rooyackers, A. S. Verhulst, S. D. Gendt, M. M. Heyns and G. Groeseneken, "Performance Enhancement in Multi-Gate Tunneling Field Effect Transistor by Scaling the Fin-Width", *Japanese Journal of Applied Physics*, vol. 49, pp. 04DC10-1–5, 2010.

[30] H. Lu and A. Seabaugh, "Tunnel FET Transistors: State-of-the-Art", *IEEE Journal of Electron Devices Society*, vol. 2, no. 4, pp. 44–49, June 2014.

[31] W. Vandenberghe, A. S. Verhulst, G. Groeseneken, B. Soree and W. Magnus, "Analytical Model for Point and Line Tunneling in a Tunnel Field-Effect Transistor", in *Proc. Int. Conf. SISPAD*, 2008, pp. 137–140.

[32] A. S. Verhulst, D. Leoneli, R. Rooyackers and G. Groeseneken, "Drain Voltage Dependent Analytical Model of Tunnel Field-Effect Transistors", *J. Appl. Phys.*, vol. 110, no. 2, pp. 024510-1–024510-10, 2011.

[33] W. G. Vandenberghe, A. S. Verhulst, K.-H. Kao, K. De Meyer and B. Sorée, "A Model Determining Optimal Doping Concentration and Material's Band Gap of Tunnel Field-Effect Transistors", *Appl. Phys. Lett.*, vol. 100, 193509 (2012).

[34] K. Gopalakrishnan, P.B. Griffin and J.D. Plummer, "I-MOS: A Novel Semiconductor Device with a Subthreshold Slope Lower than *kT/q*", in *IEDM Tech. Dig.*, 2002, pp. 289–292.

[35] C. Shen, J.-Q. Lin, E.-H. Toh, K.-F. Chang, P. Bai, C.-H. Heng, G. S. Samudra and Y.-C. Yeo, "On the Performance Limit of Impact Ionization Transistors", in *IEDM Tech. Dig.*, 2007, pp. 117–120.

[36] S. Ramaswamy and M. J. Kumar, "Junction-less Impact Ionization MOS (JIMOS): Proposal and Investigation", *IEEE Trans. on Electron Devices*, vol. 61, pp. 4295–4298, December 2014.

[37] N. Kannan and M. J. Kumar, "Dielectric-Modulated Impact-Ionization MOS (DIMOS) Transistor as a Label-Free Biosensor", *IEEE Electron Device Letters*, vol. 34, December 2013.

[38] N. Kannan and M. J. Kumar, "Charge Modulated Underlap I-MOS Transistor as a Label-Free Biosensor: A Simulation Study", *IEEE Trans. Electron Devices*, vol. 62, no. 8, pp. 2645–2651, August 2015.

[39] N. Abele, N. Fritschi, K. Boucart, F. Casset, P. Ancey and A. M. Ionescu, "Suspended-Gate MOSFET: Bringing New MEMS Functionality into Solid-State MOS Transistor", in *IEDM Tech. Dig.*, 2005, pp. 479–481.

[40] H. Kam, D. T. Lee, R. T. Howe and T.-J. King, "A New Nanoelectromechanical Field Effect Transistor (NEMFET) Design for Low-Power Electronic"s, in *IEDM Tech. Dig.*, 2005, pp. 463–466.

[41] K. Akarvardar, C. Eggimann, D. Tsamados, Y. Singh Chauhan, G. C. Wan, A. M. Ionescu, R. T. Howe and H.-S. P. Wong, "Analytical Modeling of the Suspended-Gate FET and Design Insights for Low-Power Logic", *IEEE Trans. Electron Devices*, vol. 55, no. 1, pp. 48–59, January 2008.

[42] F. Chen, H. Kam, D. Markovic, T. J. King, V. Stojanovic and E. Alon, "Integrated Circuit Design with NEM Relays", in *Proc. IEEE/ACM Int. Conf. Computer Aided Design*, 2008, pp. 750–757.

[43] K. Akarvardar, D. Elata, R. Parsa, G. C. Wan, K. Yoo, J. Provine, P. Peumans, R. T. Howe and H.-S. P. Wong, "Design Considerations for Complementary Nanoelectromechanical Logic Gates", in *IEDM Tech. Dig.*, 2007, pp. 299–302.

[44] S. Salahuddin and S. Datta, "Use of Negative Capacitance to Provide a Sub-threshold Slope Lower than 60 mV/decade", *Nanoletters*, vol. 8, no. 2, pp. 405–410, 2008.

[45] S. Salahuddin and S. Datta, "Can the Subthreshold Swing in a Classical FET be Lowered Below 60 mV/decade?" in *IEDM Tech. Dig.*, 2008, pp. 693–696.

[46] A. Padilla, C. W. Yeung, C. Shin, M. H. Cho, C. Hu and T.-J. King Liu, "Feedback FET: A Novel Transistor Exhibiting Steep Switching Behavior at Low Bias Voltages", in *IEDM Tech. Dig.*, 2008, pp. 171–174.

[47] H. Kam, T.-J. K. Liu and E. Alon, "Design Requirements for Steeply Switching Logic Devices", *IEEE Trans. on Electron Devices*, vol. 59, no. 2, February 2012.

4

Drain current modelling of tunnel FET: the task and its challenges

4.1 Introduction

In a MOSFET, free carriers from the source cross the source–channel barrier on application of drain voltage, whereas in a TFET, free carriers are generated due to band-to-band tunnelling across the source–channel junction, which are then swept towards the drain due to the applied drain electric field. As discussed in the first three chapters, since quantum mechanical effects are at the basis of the operation of a TFET, its modelling approach inherently differs from that of a MOSFET. The approaches to modelling TFETs, and quantum devices in general, can be divided into two broad categories – atomistic modelling and analytical modelling. In the case of TFETs, most atomistic modelling derives its basis from the non-quilibrium Green function (NEGF) approach. This methodology is far more accurate and powerful, especially at small device dimensions, when quantum effects become far more prominent. The first step in an NEGF approach usually involves the extraction of the band structure of the system using the density functional Theory (DFT)-based calculations. Subsequently, this band structure is used in conjunction with a ballistic quantum transport model to self-consistently solve for the

Tunnel Field-Effect Transistors (TFET): Modelling and Simulation, First Edition. Jagadesh Kumar Mamidala, Rajat Vishnoi and Pratyush Pandey.

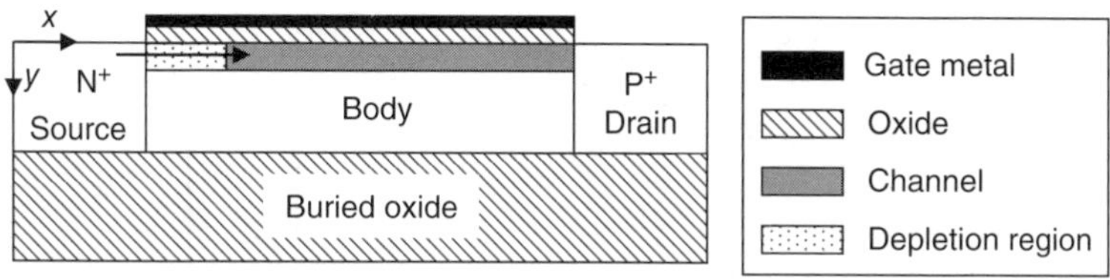

Figure 4.1 Schematic depicting the depletion region and the inversion layer charge in the channel of a TFET.

wavefunctions, allowed energy states, energy values and so on. However, this approach also has its drawbacks. First, it is computationally intensive, and running an NEGF simulation can take months on a fast cluster of computers. On the other hand, many models (such as Schenk's model) use the NEGF approach as a first step, and subsequently apply various approximations to obtain a more computationally efficient model.

The other category of modelling uses an analytical approach, where known physics equations and theories are combined, approximated, and modified to give the final results of the device. The actual theories used may vary across different approaches, ranging in complexity, from the $\vec{k} \cdot \vec{p}$ method to solving approximated versions of Poisson's equation. The advantage of the analytical approach is two-fold – it is less computationally intensive and gives a better understanding of the essential physical phenomenon behind the device functioning. This is the approach to modelling that will be explained in this book.

Figure 4.1 depicts the cross-sectional view of a TFET in the ON-state ($|V_{GS}| > V_{th}$ and $|V_{DS}| > 0$) showing the depletion region and the inversion layer charge in the channel. It shows the existence of an inversion layer (i.e. the channel) at the surface (i.e. Si–SiO_2 interface) in the body. It also shows a depletion region between the source and the channel, which is due to a difference in the type of carriers in the source and in the channel. This depletion region has a high electric field, which leads to a potential drop greater than E_g/q (where E_g is the bandgap of the material) across the source and the channel leading to the tunnelling of carriers through the bandgap. Since the inversion layer is formed at the surface, tunnelling of carriers occurs only at the surface, as shown in Figure 4.1. As can be seen in different tunnelling models discussed in Chapter 2, the generation rate of carriers for band-to-band tunnelling is a function of the electric field. Hence, to calculate the drain current in a TFET, we first need to find the electric field at the surface in the tunnelling region (i.e. the source–channel depletion region), as shown in Figure 4.1. Once the potential profile is modelled accurately, the electric field can be calculated by taking its derivative. The challenge in modelling the surface potential in the tunnelling region comes from the two-dimensional nature of the

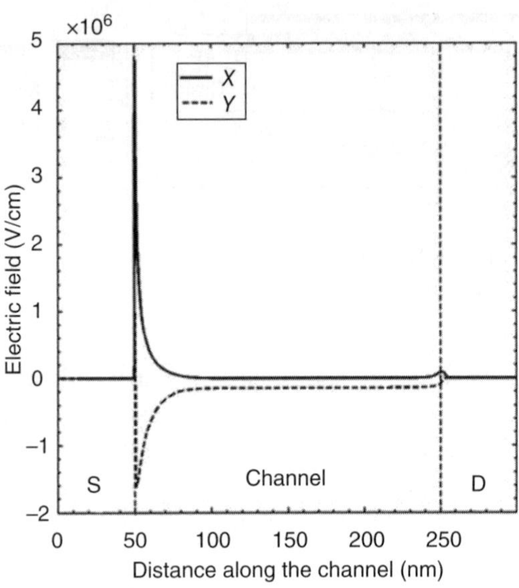

Figure 4.2 Electric field and along the surface of a TFET in the x- and the y-directions.

electric field in this region. In the channel of a MOSFET and also in that of a TFET, we take the gradual channel approximation, which assumes the electric field in the x-direction to be negligible as compared to the electric field in the y-direction, and hence we have to solve the one-dimensional Poisson equation in the channel as given below:

$$\frac{\partial^2 \psi}{\partial y^2} = \frac{q}{\varepsilon} \tag{4.1}$$

where ψ is the potential.

However, in the tunnelling region (depletion region shown in Figure 4.1), the gradual channel approximation is not valid, as can be seen in Figure 4.2 as the x- and y-direction electric fields are comparable in magnitude. Hence, we have to solve the two-dimensional Poisson equation as given below:

$$\frac{\partial^2 \psi}{\partial x^2} + \frac{\partial^2 \psi}{\partial y^2} = \frac{q}{\varepsilon} \tag{4.2}$$

Equation (4.2) cannot be integrated and solved as in the case of Equation (4.1), and hence different mathematical methods and approximations are needed to solve it.

The next step in the modelling of a TFET is finding the tunnelling generation rate using the surface potential obtained by solving Equation (4.2). The tunnelling generation rate is calculated by using local or non-local models of tunnelling, as described in Chapter 1. A local model is a function of the local electric field and gives the tunnelling generation rate at a particular point. The surface potential can be differentiated and used in a local tunnelling model. A non-local model, on the other hand, uses the tunnelling length, which is the length over which the potential has a drop of E_g/q. Hence, a non-local model becomes a function of the surface potential. The next and the most challenging step is integrating this tunnelling rate over the volume of the tunnelling region. The expression for the tunnelling generation rate G_{btb} using Kane's model (local model) is given below:

$$G_{btb} = A\frac{|E^2|}{\sqrt{E_g}}\exp\left[-B\frac{E_g^{3/2}}{|E|}\right] \tag{4.3}$$

where E is the electric field, E_g is the bandgap and A and B are tunnelling parameters.

To find the drain current I_d, we have to integrate the tunnelling generation rate over the volume of the tunnelling region:

$$I_d = q\int G_{btb}dV \tag{4.4}$$

The expression for G_{btb} in Equation (4.3) does not have a closed form integration as it involves both linear and exponential terms in the electric field. Hence, this integration is the biggest challenge in modelling the drain current in tunnel FETs and various approximations and assumptions are needed to be used to obtain an expression for the drain current. Modelling the drain current in a TFET mainly involves two steps: (i) solving for the surface potential and (ii) finding the tunnelling generation rate and integrating it over the tunnelling volume.

4.2 TFET modelling approach

In this section, using an elementary TFET model, we will describe the basic approach towards developing a drain current model for a TFET [1]. We take a p-channel SOI TFET with a high source and drain doping of $10^{21}/cm^3$, as shown in Figure 4.3. The device has a channel length of L, silicon film thickness of T_{Si}, oxide thickness T_{ox} and a body doping of N_A. The device has a long channel (i.e. channel length > 50 nm) and has a thin silicon film ($T_{Si} \sim 10$ nm) such that the body is fully depleted under the influence of a gate voltage. To model the TFET, first we have to solve for the surface potential of the device. Therefore, let us begin by observing the simulation results for the surface potential of the TFET and an equivalent MOSFET, as shown in Figure 4.4. It can be seen from this figure that

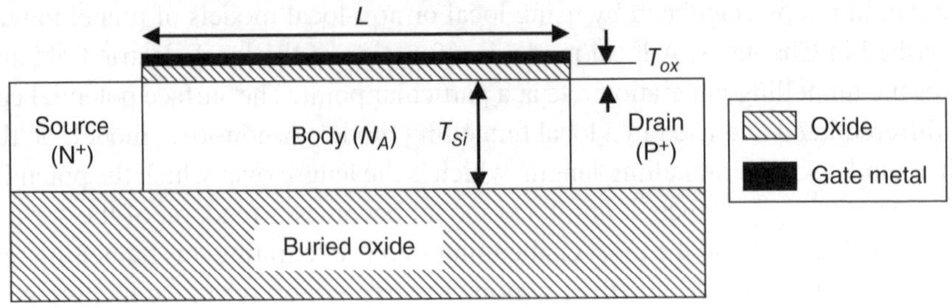

Figure 4.3 Schematic view of the TFET.

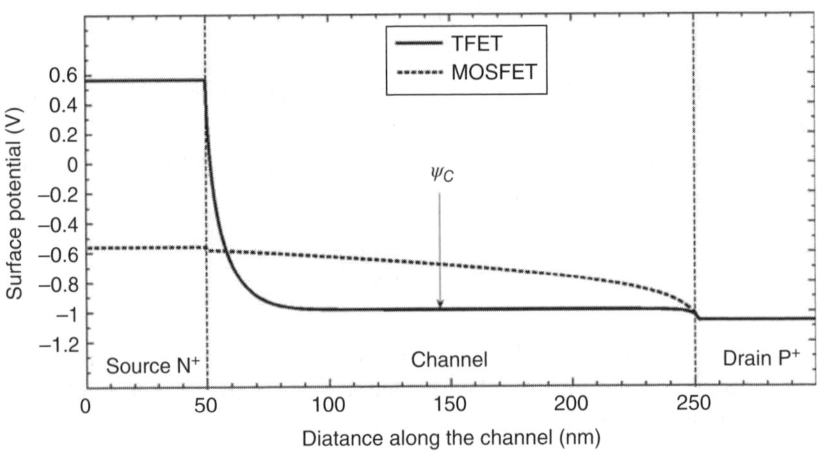

Figure 4.4 Comparison between the surface potential of the TFET shown in Figure 4.3 and that of an equivalent MOSFET.

the surface potential in the channel of the TFET is almost constant and varies rapidly in the source–channel junction. Let us assume the value of this constant potential in the channel to be ψ_C. Hence, we need to find the value of ψ_C and model the surface potential in the source channel junction.

4.2.1 Finding the value of ψ_C

As can be seen in Figure 4.4, the value of ψ_C is equal to the surface potential of the equivalent MOSFET at the drain end. Hence, we get the following expression for ψ_C:

$$\psi_C = \psi_B + V_{DS} \quad \text{for } V_{DS} < |V_{GS} - V_{Th}| \text{ in the linear region} \tag{4.5}$$

$$\psi_C = \psi_B + |V_{GS} - V_{th}| \quad \text{for } V_{DS} < |V_{GS} - V_{Th}| \text{ in the saturation region} \tag{4.6}$$

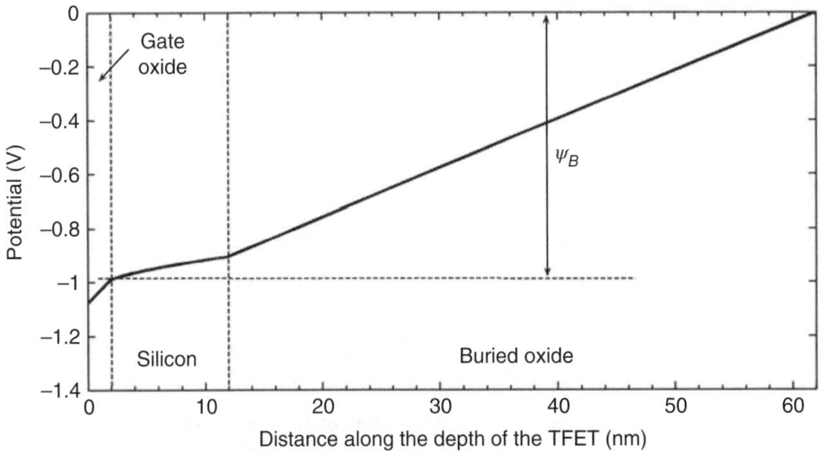

Figure 4.5 Potential profile along the y-direction of the TFET.

where ψ_B is the built-in potential of the channel, which is the sum of the band bending in the body and drop across the buried oxide (Figure 4.5).

4.2.2 Modelling the surface potential in the source–channel junction

Now we need to model the surface potential in the depletion region at the source–-channel junction (i.e. the tunnelling region), by solving the two- dimensional Poisson equation (4.3). Using a parabolic approximation for the potential in the y-direction (also known as the pseudo-2D method, which will be described in detail in the next chapter) and the y-direction boundary conditions (shown in Figure 4.6).

The potential at the Si–SiO_2 interface is equal to the surface potential:

$$\psi(x,0) = \psi_S(x) \tag{4.7}$$

The electric field at the Si-buried oxide interface is equal to zero:

$$E_y(x, T_{Si}) = 0 \tag{4.8}$$

The electric field displacement (D) is continuous across the Si–SiO_2 interface:

$$E_y(x,0) = -C_{ox}(\psi_G - \psi_S(x))/\varepsilon_{Si} \tag{4.9}$$

Using the boundary conditions given by Equations (4.7) to (4.9), we get the following general form solution for the surface potential using the two-dimensional Poisson equation:

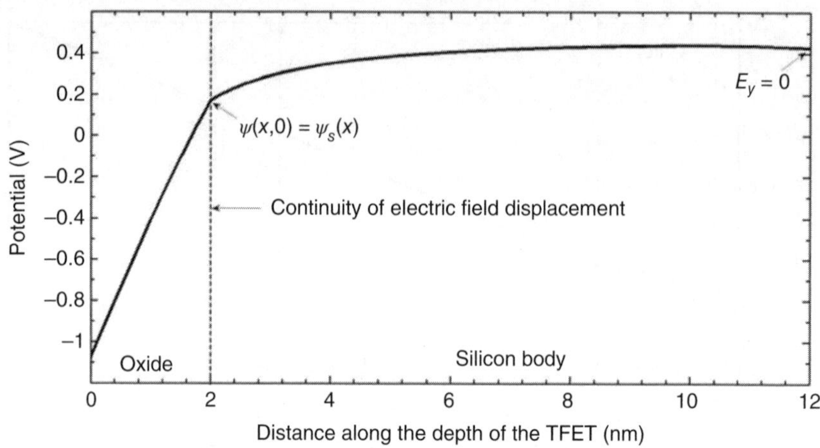

Figure 4.6 Potential profile along the y-direction in a TFET in the body region showing the boundary conditions.

$$\psi_S(x) = C\exp\left(\frac{x - x_i}{L_d}\right) + D\exp\left(\frac{-(x - x_i)}{L_d}\right) + \psi_G \tag{4.10}$$

$$L_d = \sqrt{T_{Si}T_{ax}\varepsilon_{Si}/\varepsilon_{ox}} \tag{4.11}$$

Equation (4.10) has three unknowns (C, D and x_i), which can be calculated using the boundary conditions in the x-direction (shown in Figure 4.7). Note that for modelling the surface potential in the tunnelling region, $x = 0$ is taken at the source–channel junction.

The surface potential at $x = x_i$ is equal to ψ_C:

$$\psi_S(x_i) = \psi_C \tag{4.12}$$

The electric field at $(E_x)x = x_i$ is equal to zero.

$$\frac{\partial \psi_S(x_i)}{\partial x} = 0 \tag{4.13}$$

The surface potential at the source is equal to the built-in potential between the source and the body, $V_{bi} = kt/q\ \ln\left(N_{Source}N_{body}/n_i^2\right)$, so

$$\psi_S(0) = V_{bi} \tag{4.14}$$

Using Equations (4.12) and (4.13) we get C and D:

$$C = D = (\psi_C - \psi_G)/2 \tag{4.15}$$

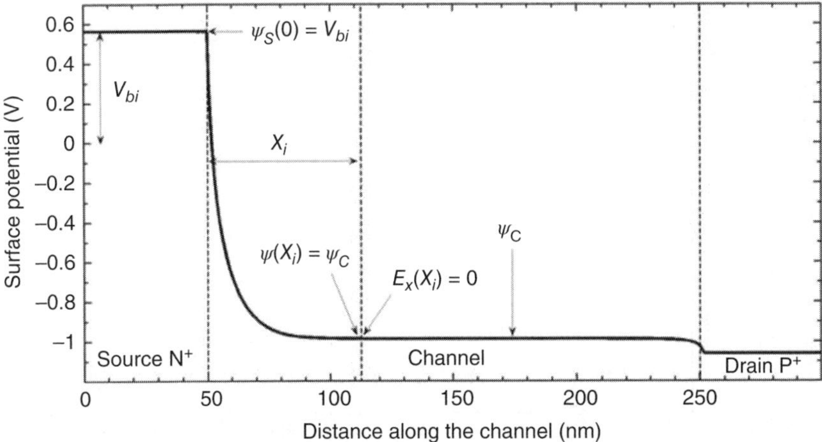

Figure 4.7 Potential profile at the surface along the x-direction of a TFET, showing the boundary conditions.

Using Equation (4.14) we get the following value of x_i:

$$x_i = L_d \cosh^{-1}[V_{bi}/(\psi_C - \psi_G)/2] \tag{4.16}$$

Finally, we get the following expression for the surface potential:

$$\psi_S(x) = (\psi_C - \psi_G)\cosh\left(\frac{x - L_d \cosh^{-1}(V_{bi}/(\psi_C - \psi_G)))}{L_d}\right) + \psi_G \tag{4.17}$$

This entire method of finding the surface potential giving us Equation (4.17) is known as the pseudo-2D model and is described in detail in the next chapter.

4.2.3 Finding the tunnelling current

Now, using the expression for surface potential given in Equation (4.17), we will find the drain current of the device. For this, we first need to find the tunnelling generation rate using Equation (4.3) and then integrate it. Figure 4.8 shows the variation in the tunnelling generation rate along the length of the device. As G_{btb} is an exponential function of the electric field, it falls sharply as we move away from the source–channel junction, where the electric field is the highest. Hence, in this model, we find this highest generation rate and multiply it by a constant factor to find the drain current. We begin with finding the minimum tunnelling length, which is the minimum length across which we get a potential drop of E_g/q, as shown in Figure 4.9 Since the curve of surface potential is steepest near the source

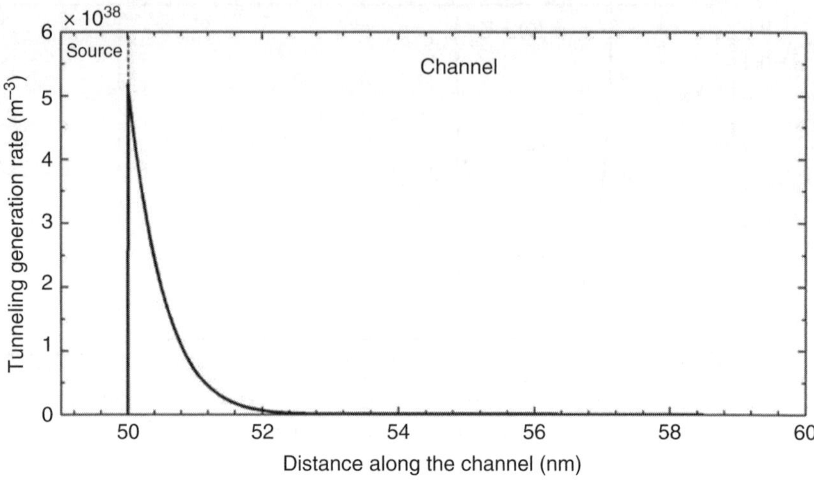

Figure 4.8 Tunnelling generation rate (G_{btb}) at the surface, along the x-direction of a TFET.

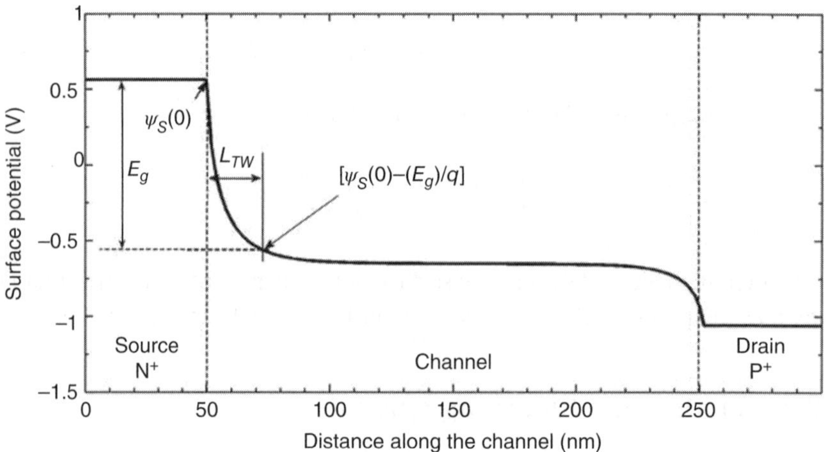

Figure 4.9 Surface potential profile at the surface, along the x-direction of a TFET, showing the shortest tunnelling length.

(as can be seen in Figure 4.6), the minimum tunnelling length (L_{TW}) will be given by the distance from the source at which the surface potential falls by E_g/q, as described by the equation below:

$$L_{TW} = x\left(\psi_S(0) - E_g/q\right) - x\left(\psi_S(0)\right) \tag{4.18}$$

Using L_{TW} we find the average electric field over the minimum tunnelling length (E_{TW}):

$$E_{TW} = E_g / qL_{TW} \tag{4.19}$$

Now we use the average electric field over the minimum tunnelling length (E_{TW}) to find the tunnelling generation rate at the source–channel interface, using Kane's model for band-to-band tunnelling (Equation (4.3)), as given below:

$$G_{btb}(0) = A\frac{|E_{TW}^2|}{\sqrt{E_g}}\exp\left[-B\frac{E_g^{3/2}}{|E_{TW}|}\right] \tag{4.20}$$

Kane's model is a local tunnelling model and the above generation rate is what we obtain by using the average electric field along the minimum tunnelling length in Kane's model. Now, since G_{btb} is an exponential function of the electric field, we assume the generation rate $G_{btb}(0)$, given by Equation (4.20) as the dominant generation rate, as it is obtained by using the highest electric field present in the device. Hence, to obtain the drain current, we multiply this generation rate ($G_{btb}(0)$) by a constant A_V, which accounts for the volume over which the generation of carriers by tunnelling is taking place (i.e. the volume of tunnelling region). This gives the following equation for the drain current:

$$I_D = A_V A\frac{|E_{TW}^2|}{\sqrt{E_g}}\exp\left[-B\frac{E_g^{3/2}}{|E_{TW}|}\right] \tag{4.21}$$

The parameter A_V has no analytical form and has to be extracted from experimental or simulated characteristics.

4.3 MOSFET modelling approach

The TFET modelling approach described in the previous sections is based on the surface potential of the device and is known as the surface potential-based modelling approach. MOSFET models, however, are based on two types of approaches. The charge-based approach and the surface potential-based approach. The basic textbook model that everybody is familiar with is a charge-based model, which gives us the following equation for the drain current I_d in the linear region [2]:

$$I_D = \frac{\mu_n C_{ox} W}{L}\left((V_{GS} - V_{th})V_{DS} - \frac{V_{DS}{}^2}{2}\right) \tag{4.22}$$

where μ_n is the electron mobility, C_{ox} is oxide capacitance, W is the channel width and L is the channel length. In a charge-based modelling approach, the channel charge density is calculated first, using the gate capacitance, and then the drain current is calculated by using the drift and diffusion equations on the channel charge.

Another approach for modelling a MOSFET is the surface potential-based approach. In this approach, we solve Poisson's equation in the channel given by the following equation and then find the charge density in the channel region using the surface potential:

$$\frac{\partial^2\psi(x,y)}{\partial x^2}+\frac{\partial^2\psi(x,y)}{\partial y^2}=\frac{q}{\varepsilon}(N_{SUB}+n(x,y)-p(x,y)) \tag{4.23}$$

where ψ is the potential, N_{SUB} is the body doping and n and p are electron and hole concentrations, respectively.

However, since the gradual channel approximation holds in the channel of a MOSFET, Equation (4.23) is simplified into a one-dimensional Poisson equation as given by

$$\frac{\partial^2\psi(x,y)}{\partial y^2}=\frac{q}{\varepsilon}(N_{SUB}+n(x,y)-p(x,y)) \tag{4.24}$$

For an n-channel MOSFET we can neglect the $p(x,y)$ term in Equation (4.24) and vice versa for a p-channel MOSFET.

After solving for the surface potential using Poisson's equation (4.24), we find the channel charge. Once the channel charge q_i is known, we apply the drift diffusion equations on the channel charge to find the drain current I_d:

$$I_d = WC_{ox}\int_0^{V_{DS}} \mu q_i dV \tag{4.25}$$

which can be rewritten in terms of surface potential ψ_S as

$$I_d = WC_{ox}\int_{x=0}^{x=L} \mu\left[\phi_i\frac{dq_i}{dx}-q_i\frac{d\psi_s}{dx}\right]dx \tag{4.26}$$

The current is dominated by diffusion of the charges in the subthreshold region and by the drift of charges in the strong inversion region.

In summary, modelling of a MOSFET using the surface potential approach involves two steps: (i) solving the simplified one-dimensional Poisson equation for finding the surface potential and (ii) using the surface potential for calculating the inversion charge and then using the drift diffusion equation for finding the drain current.

References

[1] J. Wan, C. L. Royer, A. Zaslavsky and S. Cristoloveanu, "A tunneling field effect transistor model combining interband tunneling with channel transport", *J. Appl. Phys.*, vol. 110, no. 10, pp. 104503–104503-7, 2011.

[2] B.G. Streetman and S.K. Banerjee, *Solid State Electronic Devices*, 6th edn, Pearson Education, 2006.

5

Modelling the surface potential in TFETs

By now you are familiar with the basic approach of modelling a TFET, as shown in Figure 5.1. In the previous chapter, we identified the two main steps in the modelling of TFETs. In this chapter, we will discuss the first step, that is solving for the surface potential, in detail. In the next chapter, we will be covering the second step, that is finding the tunnelling generation rate and integrating it over the tunnelling volume.

To model the surface potential, let us write the two-dimensional Poisson equation in the channel of the TFET:

$$\frac{\partial^2\psi(x,y)}{\partial x^2}+\frac{\partial^2\psi(x,y)}{\partial y^2}=\frac{qN_A}{\varepsilon_{Si}} \tag{5.1}$$

where N_A is the body doping of the TFET. Equation (5.1) does not include the mobile inversion charge in the channel. The solution to Equation (5.1) will give us the surface potential. Equation (5.1) is a second-order differential equation in two variables and, hence, it cannot be integrated in a straightforward way. Solving this equation is a major challenge in modelling a TFET. Hence, in this chapter, we describe various approaches and methods that deal with solving Equation (5.1) and finding the surface potential in a TFET. These approaches involve various approximations, assumptions, simplifications and mathematical techniques to solve the two-dimensional (2D) Poisson equation.

Tunnel Field-Effect Transistors (TFET): Modelling and Simulation, First Edition. Jagadesh Kumar Mamidala, Rajat Vishnoi and Pratyush Pandey.

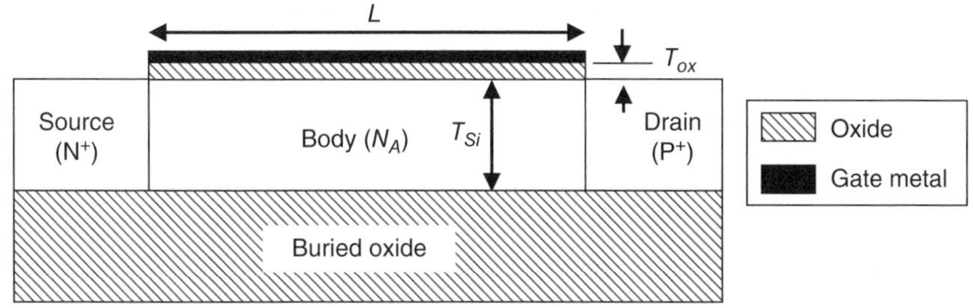

Figure 5.1 Schematic view of a TFET.

5.1 The pseudo-2D method

The Pseudo-2D method, also known as the parabolic approximation method, is one of the most widely used methods for modelling the surface potential in a TFET. This method essentially simplifies the 2D Poisson equation (5.1) into a second-order 1D linear differential equation, which is then solved using basic mathematical techniques. The method was first suggested by K. K. Young [1] and has since then become popular for modelling TFETs and for studying short channel effects (SCEs) in MOSFETs [2, 3]. In this section, we will describe the pseudo-2D method for solving the surface potential of a TFET [4–19].

5.1.1 Parabolic approximation of potential distribution

Let us first take a look at the potential distribution in a TFET in the x- and y-directions as given in Figures 5.2 and 5.3, respectively. As can be seen in Figure 5.3, the potential distribution at any $x = x_i$ along the y-direction (i.e. $\psi(x, y)$) is monotonous in nature and can be approximated by a polynomial function, as given below:

$$\psi(x_i, y) = a_0 + a_1 y + a_2 y^2 + a_3 y^3 \tag{5.2}$$

where a_0, a_1, a_2, etc., are coefficients of the polynomial in y. At each value of x (i.e. $x = x_i$), we have a different set of polynomial coefficients $a_0(x), a_1(x), a_2(x)$, etc. Hence, Equation (5.2) becomes

$$\psi(x, y) = a_0(x) + a_1(x)y + a_2(x)y^2 + a_3(x)y^3 \tag{5.3}$$

To evaluate $\psi(x, y)$, we have to first find the polynomial coefficients ($a_0(x)$, $a_1(x), a_2(x), \ldots$). The coefficients can be found by making use of the following boundary conditions in the y-direction.

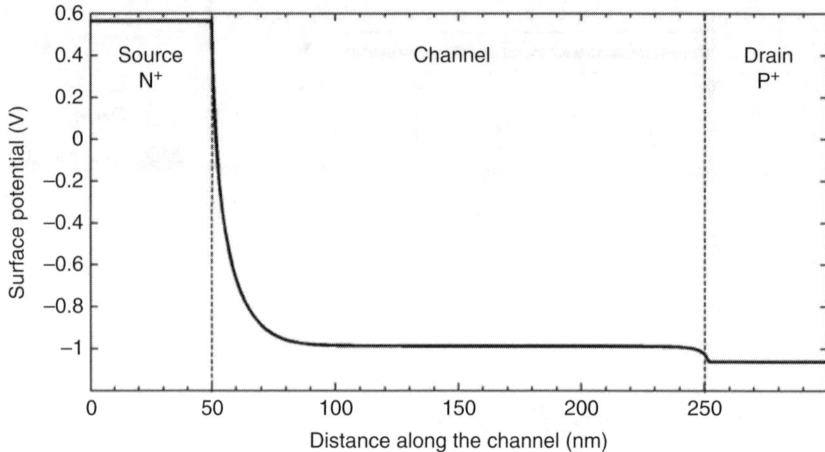

Figure 5.2 Potential distribution along the x-direction of a TFET at the surface.

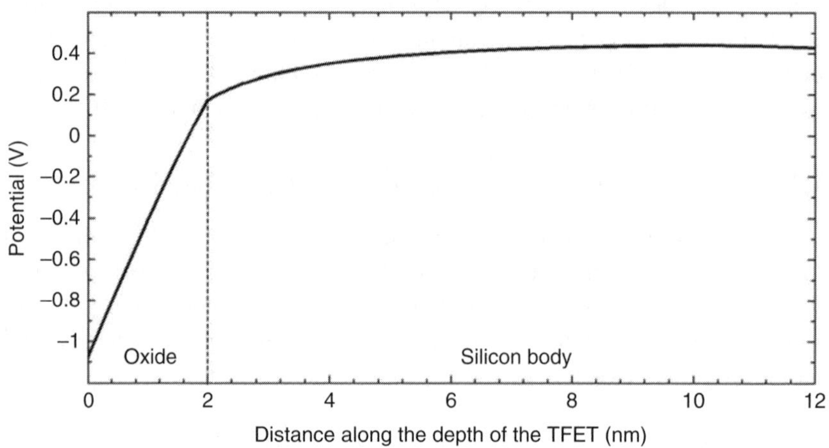

Figure 5.3 Potential distribution along the y-direction of a TFET.

(i) The electric field at the Si-buried oxide interface (i.e. $x = t_{Si}$) is equal to zero, which gives us

$$\left.\frac{\partial \psi(x,y)}{\partial y}\right|_{y=t_{Si}} = 0 \tag{5.4}$$

(ii) The potential at the $Si-SiO_2$ interface (i.e. $x=0$) is equal to the surface potential ($\psi_S(x)$), which gives us:

$$\psi(x,0)=\psi_S(x) \tag{5.5}$$

(iii) The electric field displacement is continuous across the $Si-SiO_2$ interface (i.e. $x=0$), which gives us

$$\left.\frac{\partial\psi(x,y)}{\partial y}\right|_{y=0}=\frac{-C_{ox}(\psi_G-\psi_S(x))}{\varepsilon_{Si}} \tag{5.6}$$

where C_{ox} is the gate oxide capacitance per unit area and ψ_G is the gate potential. The gate potential ψ_G is given by

$$\psi_G=V_{GS}-V_{FB} \tag{5.7}$$

where V_{FB} is the flat-band voltage and is dependent on the work function of the gate.

Finally, there are three boundary conditions known to us, in the y-direction, as given by Equations (5.4) to (5.6). Using these boundary conditions, we can only evaluate the first three polynomial coefficients in Equation (5.3). We have to ignore the higher-order terms (i.e. terms y^3, y^4, y^5 and so on), which gives us the following form of Equation (5.3):

$$\psi(x,y)=a_0(x)+a_1(x)y+a_2(x)y^2 \tag{5.8}$$

Hence, this method for solving the surface potential is known as the parabolic approximation approach. Here we are approximating the potential distribution in the y-direction by using a parabolic (i.e. a second-order polynomial) expression (5.8). We will now find the coefficients $a_0(x)$, $a_1(x)$ and $a_2(x)$ using the boundary conditions (5.4) to (5.6).

Substituting Equation (5.8) into Equation (5.4) we get

$$a_1(x)y+2a_2(x)t_{Si}=0 \tag{5.9}$$

Substituting Equation (5.8) into Equation (5.5) we get

$$a_0(x)=\psi_S(x) \tag{5.10}$$

Finally, substituting Equation (5.8) into Equation (5.6) we get

$$a_1(x)=\frac{-C_{ox}(\psi_G-\psi_S(x))}{\varepsilon_{Si}} \tag{5.11}$$

Equations (5.10) and (5.11) give us $a_0(x)$ and $a_1(x)$, respectively. Using Equation (5.11) in (5.9) gives us $a_2(x)$.

$$a_2(x) = \frac{C_{ox}(\psi_G - \psi_S(x))}{2t_{Si}\varepsilon_{Si}} \tag{5.12}$$

Now, by using Equations (5.10) to (5.12), Equation (5.8) becomes

$$\psi(x,y) = \psi_S(x) + \frac{-C_{ox}(\psi_G - \psi_S(x))}{\varepsilon_{Si}}y + \frac{C_{ox}(\psi_G - \psi_S(x))}{2t_{Si}\varepsilon_{Si}}y^2 \tag{5.13}$$

Equation (5.13) gives the complete form of the parabolic approximation of potential in the y-direction.

5.1.2 Solving the 2D Poisson equation using parabolic approximation

Using Equation (5.13) in Equation (5.1) and solving at $y=0$ simplifies the 2D Poisson equation (5.1) into the following second-order 1D linear differential equation in the surface potential, as given below:

$$\frac{\partial^2\psi_S(x)}{\partial x^2} - \frac{C_{ox}}{t_{Si}\varepsilon_{Si}}\psi_S(x) = \frac{qN_A}{\varepsilon_{Si}} - \frac{C_{ox}}{t_{Si}\varepsilon_{Si}}\psi_G \tag{5.14}$$

The homogenous part of Equation (5.14) is

$$\frac{\partial^2\psi_S(x)}{\partial x^2} - \frac{C_{ox}}{t_{Si}\varepsilon_{Si}}\psi_S(x) = 0 \tag{5.15}$$

The general form solution of Equation (5.15) is

$$\psi_S(x) = C\exp\left(\frac{x}{L_d}\right) + D\exp\left(\frac{-x}{L_d}\right) \tag{5.16}$$

where the characteristic length is

$$L_d = \sqrt{t_{Si}\varepsilon_{Si}/C_{ox}} = \sqrt{t_{Si}t_{ox}\varepsilon_{Si}/\varepsilon_{ox}} \tag{5.17}$$

A particular solution of Equation (5.14) is

$$\psi_S(x) = \psi_G - \frac{qN_AL_d^2}{\varepsilon_{Si}} \tag{5.18}$$

We get the final solution to Equation (5.14) by adding the homogenous solution and the particular solution, which is given by

$$\psi_S(x) = C\exp\left(\frac{x}{L_d}\right) + D\exp\left(\frac{-x}{L_d}\right) + \psi_G - \frac{qN_S L_d^2}{\varepsilon_{Si}} \tag{5.19}$$

Equation (5.19) gives us the general form solution for the surface potential of a TFET, but our task is not yet over. Equation (5.19) has two unknown coefficients (C and D). To find these coefficients, we make use of the x-direction boundary conditions.

5.1.3 Solution for the surface potential

Let us look at the surface potential distribution of a TFET as given in Figure 5.4. We can divide the body of the TFET into two regions, region R1 (the tunnelling region) and region R2 (the channel region), as shown in Figure 5.4. As can be seen, in the channel region the surface potential is almost constant. This is because the x-direction electric field in the tunnelling region is high and the x-direction electric field in the channel region is small. Let us define this constant potential as ψ_C. The value of ψ_C can be calculated by using the similarity in the drain sides of a MOSFET and a TFET. As can be seen in Figure 5.4, the value of ψ_C is equal to the surface potential of the equivalent MOSFET at the drain end, which gives the following expressions for ψ_C:

$$\psi_C = \psi_B + V_{DS} \quad \text{for} \quad V_{DS} < |V_{GS} - V_{Th}| \quad \text{in the linear region} \tag{5.20}$$

$$\psi_C = \psi_B + |V_{GS} - V_{th}| \quad \text{for} \quad V_{DS} > |V_{GS} - V_{Th}| \quad \text{in the saturation region} \tag{5.21}$$

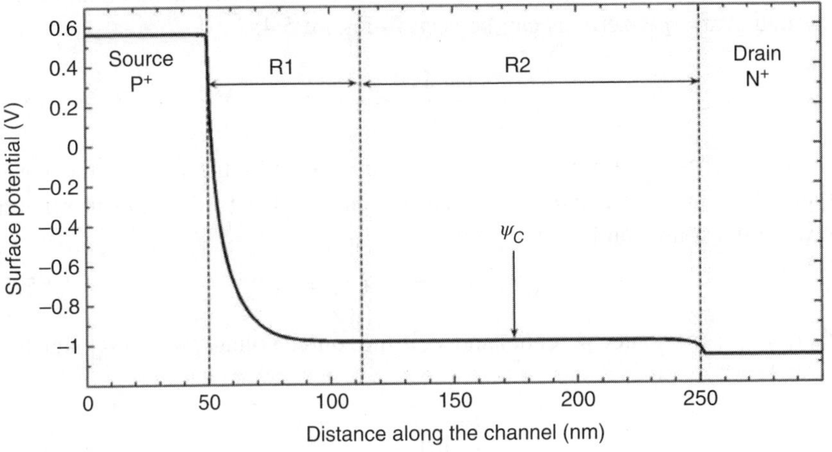

Figure 5.4 Surface potential along the channel of a TFET.

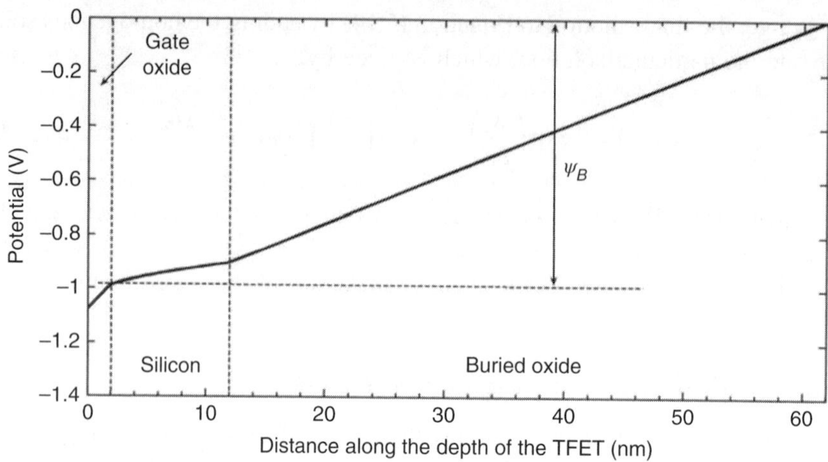

Figure 5.5 Potential along the depth (y-direction) of a TFET.

where ψ_B is the built-in potential of the channel, which is the sum of the band bending in the body and drop across the buried oxide (Figure 5.5) and V_{th} is the threshold voltage of a MOSFET of an equivalent structure. This gives us the value of the surface potential in region R2. Also, the value of the surface potential at the boundary of region R1 and region R2 is ψ_C, which is our first x-direction boundary condition:

$$\psi_S(x_i) = \psi_C \tag{5.22}$$

where x_i is the position of the boundary between the region R1 and region R2 (also the length of region R1), the value of which is unknown to us at this point.

Our second x-direction boundary condition comes from the fact that the electric field at $x = x_i$ is zero, as can be seen in Figure 5.4:

$$\frac{\partial \psi_S(x_i)}{\partial x} = 0 \tag{5.23}$$

The third x-direction boundary condition is given by the value of the surface potential at the source–body interface, which is equal to the built-in potential between the source and the body (V_{bi}):

$$\psi_S(0) = \psi_{src} = V_S + V_{bi} = V_{bi} \tag{5.24}$$

where ψ_{src} is the source potential and V_S is the source voltage, which is taken to be zero. Here

$$V_{bi} = \frac{kT}{q} \ln\left(\frac{N_{Source} N_A}{n_i^2}\right) \tag{5.25}$$

Equations (5.22) to (5.24) give us three x-direction boundary conditions. Now, using these three boundary conditions, we will find the three unknowns (i.e. C, D and x_i), but before that we will make a small change in the form of Equation (5.19), as given below:

$$\psi_S(x) = C\exp\left(\frac{(x-x_i)}{L_d}\right) + D\exp\left(\frac{(x-x_i)}{L_d}\right) + \psi_G - \frac{qN_S L_d^2}{\varepsilon_{Si}} \tag{5.26}$$

In Equation (5.26), we have shifted the centre of the exponential to $x = x_i$, which will help us to obtain a simpler solution for the surface potential. Please note that the general form of Equation (5.26) is the same as that of Equation (5.19). Only the expressions of the coefficients C and D would be simpler for Equation (5.26) as compared to Equation (5.19). This will become clear later, as we proceed with the solution.

Now, using Equation (5.26) in Equation (5.23) will give us

$$C - D = 0 \tag{5.27}$$

$$\Rightarrow C = D \tag{5.28}$$

Hence, Equation (4.26) can be rewritten as

$$\psi_S(x) = C\cosh\left(\frac{(x-x_i)}{L_d}\right) + \psi_G - \frac{qN_S L_d^2}{\varepsilon_{Si}} \tag{5.29}$$

The above simplification would not have been possible if we had not shifted the centre of the exponential of Equation (5.19) to $x = x_i$.

Using Equation (5.29) in Equation (5.22) gives us

$$C = \left(\psi_C - \left(\psi_G - \frac{qN_A L_d^2}{\varepsilon_{Si}}\right)\right) \tag{5.30}$$

Typically the body doping N_A is very low in a TFET and, hence, the term $qN_A L_d^2/\varepsilon_{Si}$ in Equation (5.30) becomes negligible as compared to the other terms. This gives us

$$C = (\psi_C - \psi_G) \tag{5.31}$$

Using Equation (5.29) in Equation (5.24) we get

$$x_i = L_d \cosh^{-1}[V_{bi}/(\psi_C - \psi_G)] \tag{5.32}$$

Finally, we have the values of all three unknown parameters (C, D and x_i). We can now write the final expression for the surface potential of the TFET, as follows:

$$\psi_S(x) = (\psi_C - \psi_G)\cosh\left(\frac{x - \left(L_d \cosh^{-1}(V_{bi}/\psi_C - \psi_G)\right)}{L_d}\right) + \psi_G \quad \text{in region R1} \tag{5.33}$$

$$\psi_S(x) = \psi_C \quad \text{in region R2} \tag{5.34}$$

Equations (5.33) and (5.34) give the final expression for the surface potential of a TFET using the pseudo-2D model. It includes the dependence of gate and drain bias voltages through the factors ψ_G and ψ_C, respectively.

The approach that we have discussed in this section does not include the effect of inversion charges formed in the channel region at high V_{GS}. We will now discuss a different method, known as the variational approach, which modifies the results obtained in this section to include the effect of inversion charges on the surface potential of a TFET.

5.2 The variational approach

The variational approach is a method for modelling the surface potential of a TFET, which incorporates the effect of the inversion charge unlike the pseudo-2D approach. It uses the calculus of variations for solving the 2D Poisson equation, which transforms the process of solving the 2D Poisson equation (5.1) into the process of finding the extremum of the Lagrangian function [20, 21] given below:

$$L_\psi = \iint \left(\frac{\varepsilon}{2}\left[\left(\frac{\partial\psi}{\partial x}\right)^2 + \left(\frac{\partial\psi}{\partial y}\right)^2\right] - \rho\psi\right) dxdy \tag{5.35}$$

which gives us the solution for ψ.

This method starts with the basic solution for the surface potential obtained in Section 5.1.3 and incorporates the effect of the inversion charge on the surface potential. This is done by modelling the change in the characteristic length (L_d) (5.17) by considering its dependence on V_{GS}. We will now describe this method in detail.

5.2.1 The variational form of Poisson's equation

In this section, we will illustrate how the problem of solving the 2D Poisson equation gets transformed into finding the extremum of the Lagrangian (5.35), using the calculus of variations and vector calculus.

Let us consider a differential equation of the following form:

$$Au=f;\ (x,y)\in D \tag{5.36}$$

where A is an operator, u is a variable, f is a function and D is the solution space.

The theory of calculus of variations states that [21] if the differential equation (5.36) has a solution for u, then it corresponds to finding the extremum (minimum or maximum) of the functional $I(u)$:

$$I(u)=(Au,u)+2(u,f) \tag{5.37a}$$

where the parenthesis operator is the scalar product of the two quantities defined as in the solution space of the functional $I(u)$. For two general functions f and g, this operator is defined as

$$(f,g)=\iint_D fg\,dx\,dy \tag{5.37b}$$

Now let us consider the 2D Poisson equation:

$$\frac{\partial^2\psi(x,y)}{\partial x^2}+\frac{\partial^2\psi(x,y)}{\partial y^2}=\frac{-\rho(x,y)}{\varepsilon};\ (x,y)\in D \tag{5.38}$$

In this case we have

$$A=\frac{\partial^2}{\partial x^2}+\frac{\partial^2}{\partial y^2} \tag{5.39a}$$

$$u=\psi \tag{5.39b}$$

$$f=-\frac{\rho}{\epsilon} \tag{5.39c}$$

Therefore,

$$I(u)=\iint\psi\left(\frac{\partial^2\psi}{\partial x^2}+\frac{\partial^2\psi}{\partial y^2}\right)dxdy+2\iint\frac{\psi\rho}{\epsilon}dxdy=I_1+I_2 \tag{5.40}$$

Let us now use the 2D divergence theorem to simplify the integral I_1. The 2D divergence theorem states that

$$\oint F_y dx - F_x dy = \iint \frac{\partial F_x}{\partial x} + \frac{\partial F_y}{\partial y} dxdy \tag{5.41}$$

where the contour of integration on the left-hand side is the surface of the domain where the problem is being solved. If we choose then we get

$$F_x = \psi \frac{\partial \psi}{\partial y}; F_y = \psi \frac{\partial \psi}{\partial y} \tag{5.42}$$

$$\frac{\partial F_x}{\partial x} = \left(\frac{\partial \psi^2}{\partial x^2}\right) + \psi \frac{\partial^2 \psi}{\partial x^2} \tag{5.43a}$$

$$\frac{\partial F_y}{\partial y} = \left(\frac{\partial \psi^2}{\partial y^2}\right) + \psi \frac{\partial^2 \psi}{\partial y^2} \tag{5.43b}$$

Substituting these values on the right-hand side of the 2D divergence theorem, we can write

$$\iint \psi \left(\frac{\partial^2 \psi}{\partial x^2} + \frac{\partial^2 \psi}{\partial y^2}\right) dxdy = \oint \left(\psi \frac{\partial \psi}{\partial y} dx - \psi \frac{\partial \psi}{\partial y} dy\right) - \iint \left(\left(\frac{\partial \psi}{\partial x}\right)^2 + \left(\frac{\partial \psi}{\partial y}\right)^2\right) dx\, dy \tag{5.44}$$

If we impose the boundary condition that, on the surface of our domain,

$$\frac{\partial \psi}{\partial y} = 0 = \frac{\partial \psi}{\partial x} \tag{5.45}$$

then the contour integral in Equation (5.44) reduces to zero. In this case, we get

$$I_1 = -\iint \left(\left(\frac{\partial \psi}{\partial x}\right)^2 + \left(\frac{\partial \psi}{\partial y}\right)^2\right) dx\, dy \tag{5.46}$$

which, if substituted in Equation (5.40), gives

$$I(u) = \iint \left[2\frac{\rho\psi}{\epsilon} - \left[\left(\frac{\partial \psi}{\partial x}\right)^2 + \left(\frac{\partial \psi}{\partial y}\right)^2\right]\right]$$

$$dx\, dy = \iint \left[2\frac{\rho\psi(x,y)}{\epsilon} - \left[\left(\frac{\partial \psi(x,y)}{\partial x}\right)^2 + \left(\frac{\partial \psi(x,y)}{\partial y}\right)^2\right]\right] dx\, dy \tag{5.47}$$

Finding the extremum of the integral in Equation (5.47) gives us the solution for $\psi(x, y)$. Hence, solving the 2D Poisson equation becomes equivalent to finding the extremum of Equation (5.47). Now we will use the variational form of Poisson's equation (5.47) to solve for the surface potential of a TFET.

5.2.2 Solution of the variational form of Poisson's equation in a TFET

In this section, we will solve the surface potential of a TFET using the variational form of Poisson's equation (5.47) [20]. Let us consider an n-channel SOI TFET as shown in Figure 5.6. The device has a channel length of L, silicon film thickness of T_{Si}, oxide thickness T_{ox} and a body doping of N_A. The device has a thin silicon film ($T_{Si} \sim 10\,\text{nm}$) such that the body is fully depleted under the influence of the gate voltage. As seen in the previous section, a TFET has two regions, the tunnelling region R1 and the channel region R2. Also, in Section 5.1.2, we have seen that the surface potential in a TFET in the tunnelling region R1 can be expressed in the following form:

$$\psi(x,0) = \psi_C + (\psi_{src} - \psi_C)e^{-\frac{x}{\lambda}} \tag{5.48}$$

where

$$\text{Characteristic length } L_d = \lambda = \sqrt{t_{Si}\varepsilon_{Si}/C_{ox}} = \sqrt{t_{Si}t_{ox}\varepsilon_{Si}/\varepsilon_{ox}} \text{ and } \psi_{src} = \text{source potential} \tag{5.49}$$

According to Equation (5.49), λ is not a function of V_{GS}. However, as we shall discuss now, this is not valid when an inversion layer is formed in the channel. To understand this, let us observe the surface potential variation with increasing V_{GS} at a fixed V_{DS}, as shown in Figure 5.7 We can observe that the surface potential in

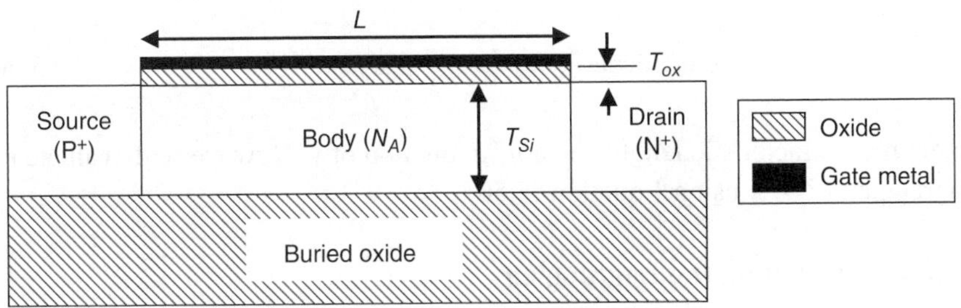

Figure 5.6 Schematic of an n-channel TFET.

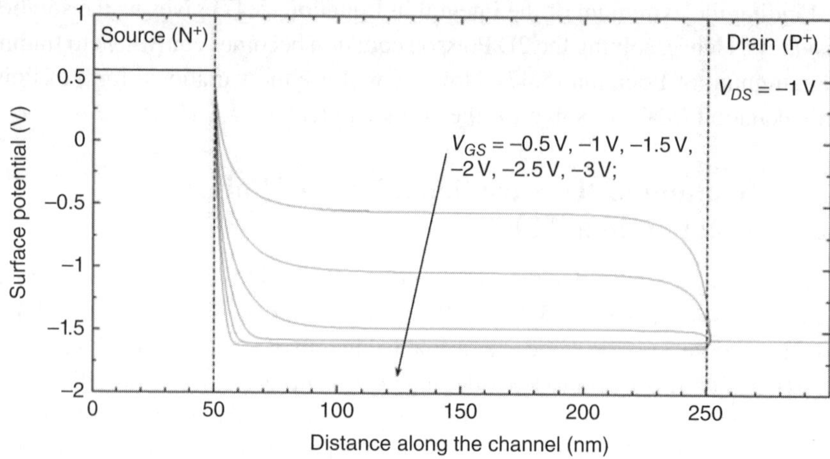

Figure 5.7 Surface potential along the channel of the TFET for different gate voltages.

the channel (region R2) becomes pinned after a certain threshold V_{GS}, at which the inversion layer is formed. Hence, using Equation (5.48), the electric field in the x-direction (E_x) in region R1 should become constant after the formation of the inversion layer. However, as can be seen in Figure 5.7, even after the formation of the inversion layer, E_x in region R1 continues to increase. This is possible only if the factor λ decreases with increasing V_{GS}. Hence, it becomes important to incorporate the effect of V_{GS} on λ for accurately modelling the surface potential in region R1. Note that in the pseudo-2D method, λ or L_d was independent of V_{GS}, as can be seen in Equation (5.17). Figure 5.8 shows the potential profile along the x-direction for different values of y in the body of the TFET. Here we can observe that the potential profile for any value of y can be expressed by an equation similar to (5.48). Hence, we can write:

$$\psi(x,y) = \psi_C(y) + (\psi_{src} - \psi_C(y))e^{-x/\lambda(y)} \tag{5.50}$$

where the characteristic length λ is now a function of y. Now we will evaluate the Lagrangian (5.47), using Equation (5.50).

5.2.2.1 Evaluating the Lagrangian

In this section, we will solve for $\psi(x, y)$ using Equation (5.50) in the Lagrangian for Poisson's equation (5.47) and minimise it. However, let us first make a

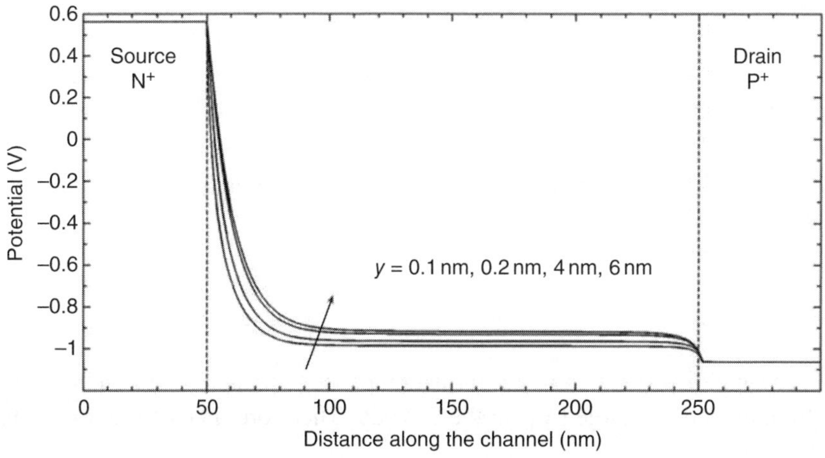

Figure 5.8 Potential along the channel of the TFET at different depths (y) *from the surface.*

simplification in Equation (5.50). As Poisson's equation is linear, using the principle of superposition, let us define the following two parameters:

$$\phi(x,y) = \psi(x,y) - \psi_C(y) \tag{5.51}$$

and

$$P(x,y) = \rho(x,y) - N_A \tag{5.52a}$$

Using Equation (5.52a) we have subtracted the effect of depletion charge (N_A) in Poisson's equation (5.38). Equation (4.51) gives us a simplified form of Equation (5.50):

$$\phi(x,y) = (\psi_{src} - \psi_C(y))e^{-x/\lambda(y)} \tag{5.52b}$$

Note that our ultimate aim is to find the tunnelling current, for which we have to find the x-direction electric field (E_x), and note that $\partial\phi(x,y)/\partial x$ in Equation (5.52b) is equal to $\partial\psi(x,y)/\partial x$ in Equation (5.50). Hence, we get the same E_x using Equations (5.51) and (5.52b). Substituting Equation (5.52b) in Equation (5.47), we get the first term of the Lagrangian (5.47) as

$$\begin{aligned} L_{\psi 1} &= \iint_D \frac{\varepsilon}{2}\left[\left(\frac{\partial\phi(x,y)}{\partial x}\right)^2 + \left(\frac{\partial\phi(x,y)}{\partial y}\right)^2\right] dx\,dy \\ &= \int \frac{\varepsilon}{4}\left\{\frac{1+0.5(\lambda'(y))^2}{\lambda(y)}\psi_0^2(y) + \left(\psi_0(y)'\right)^2\lambda + \psi_0(y)\psi_0(y)'\lambda'\right\} dy \end{aligned} \tag{5.53}$$

where

$$\psi_0(y)=\psi(x=0)=\psi_{scr}(y)-\psi_c(y);\ \psi_0'(y)=\frac{\partial\psi_0(y)}{\partial y}\ \text{and}\ \lambda'(y)=\frac{\partial\lambda(y)}{\partial y}$$

Next we will calculate the second term of the Lagrangian (5.47):

$$L_{\psi 2}=\iint_D [\phi(x,y)P(x,\ y)]dx\,dy \tag{5.54}$$

For small V_{GS} (i.e. before the formation of the inversion layer), there exists a uniform depletion charge (N_A) in the Si body. Therefore, $P(x, y)$ is zero, leading to $L_{\psi 2}=0$.

However, when the gate voltage (V_{GS}) increases, an inversion layer forms at the surface in the channel, which makes $P(x, y)$ non-zero. Using the charge sheet approximation (i.e. $P(x, y)$ is non-zero only at the surface $y=0$) and assuming the electron quasi-Fermi level to be uniform in the x-direction, we get

$$L_{\psi 2}=-\int_0^{\infty}\int_0^{T_{si}}\psi(x,y)P(x,y)dy\,dx=-\int_0^{\infty} qN_{inv}\left(1-e^{q/kT\psi(x,0)}\right)\psi(x,0)\,dx \tag{5.55}$$

using the fact that, for a potential distribution given in Equation (5.50),

$$\frac{\partial\psi}{\partial x}=-\frac{\psi}{\lambda} \tag{5.56}$$

we get

$$L_{\psi 2}=qN_{inv}\lambda(0)\int_{\psi_0(0)}^{0}\left(1-e^{q\psi/kT}\right)d\psi\approx qN_{inv}\lambda(0)[-\psi_0(0)] \tag{5.57}$$

where $\psi_0(y)=\psi(0,y)$.

Now adding Equations (5.47) and (5.57) we get

$$L_{\psi}\int_{-T_{ox}}^{T_{Si}}\frac{\varepsilon}{4}\left\{\frac{1+0.5(\lambda'(y))^2}{\lambda(y)}\psi_0^2(y)+\left(\psi_0(y)'\right)^2\lambda+\psi_0(y)\psi_0(y)'\lambda'\right\}dy \tag{5.58}$$
$$+qN_{inv}\lambda(0)[-\psi_0(0)]$$

5.2.2.2 Minimising the Lagrangian

The integrant equation (5.58) cannot be analytically integrated unless some assumptions are made. We drop the $\psi_0(y)'$ term in the silicon body because the vertical electric field in the Si body is small. Thus, we integrate the term $(\psi_0(y)')^2\lambda$ only in the oxide region. Second, we drop the term $\psi_0(y)\psi_0(y)'\lambda'$ in both the silicon and the oxide regions as λ is not expected to change much in the thin gate oxide. Finally, we make an assumption that λ is constant, giving $\lambda'=0$. This assumption is in contradiction with our original premise that λ is a function of y as given in Equation (5.50). However, for the time being let us use this assumption, which gives us the following expression for L_ψ:

$$L_\psi = \frac{\varepsilon_{Si}}{4}\frac{T_{Si}}{\lambda}\psi_0^2(0) + \frac{\varepsilon_{ox}}{4}\frac{\lambda}{T_{ox}}\psi_0^2(0) - qN_{inv}\lambda\psi_0(0) \tag{5.59}$$

Solving for $dL_\psi/d\lambda = 0$, we obtain

$$\frac{1}{\lambda^2} = \frac{\varepsilon_{ox}}{\varepsilon_{Si}T_{Si}T_{ox}} - \frac{4qN_{inv}}{\varepsilon_{Si}T_{Si}\psi_0(0)} \tag{5.60}$$

Equation (5.60) shows that λ decreases with an increase in N_{inv} (as $\psi_0(0)<0$). From Equation (5.60), we can observe that in the absence of the inversion layer (i.e. when V_{GS} is below the threshold), $N_{inv}=0$, which gives us the same expression for λ as that for L_d in Equation (5.49). When N_{inv} is very large, asymptotically we have $\lambda \sim N_{inv}^{-0.5}$ from Equation (5.60).

5.2.2.3 Empirical formulation of characteristic length

However, in the presence of the inversion charge, it is not a good practice to assume λ to be constant in the y-direction. This is because the inversion charge is only present at the surface and λ reduces aggressively at the surface and remains constant deep inside the Si body. Therefore $\lambda(y=0)$ or $\lambda(0)$ should reduce faster with N_{inv} than predicted by Equation (5.60). Hence, we will adopt an empirical form for $\lambda(y)$ based on observations made from TCAD simulations.

Figure 5.9 shows the TCAD simulation results for variation in $\lambda(y)$ with y for different values of T_{Si}. By observing these curves, we adopt the following empirical expression for $\lambda(y)$:

$$\lambda(y) = \Lambda \sin\left(\frac{\pi}{2}\frac{y+y_0}{T_{Si}+y_0}\right) \tag{5.61}$$

where y_0 is a constant parameter. Equation (5.61) in Equation (5.58) gives

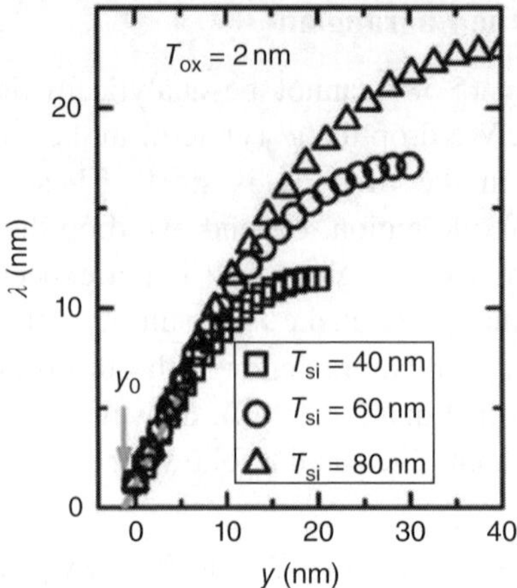

Figure 5.9 Variation in λ(y) with y for different values of T_{Si} [20].

$$L_{\psi} = \frac{\varepsilon_{Si}}{4}\psi_0^2(0)\int_0^{T_{Si}} \frac{1+0.5(\lambda')^2}{\lambda}dy + \frac{\varepsilon_{ox}}{4}\frac{\lambda(0)}{T_{ox}}\psi_0^2(0) - qN_{inv}\lambda(0)\psi_0(0) \tag{5.62}$$

Now we take (another empirical formulation)

$$\Lambda = \frac{2}{\pi}(T_{Si}+y_0) \tag{5.63}$$

which gives

$$A = \int_0^{T_{Si}} \frac{1+0.5(\lambda')^2}{\lambda}dy = \frac{3}{2}\ln\left(\cot\frac{\pi}{4}\frac{y_0}{T_{Si}+y_0}\right) - \frac{1}{2}\cos\frac{\pi}{2}\frac{y_0}{T_{Si}+y_0} \tag{5.64}$$

Using the Taylor series expansion of Equation (5.64) and using the most significant term, we can show that

$$\frac{dA}{dy_0} \approx \frac{3}{4}\frac{T_{Si}}{y_0}\frac{1}{T_{Si}+y_0} \tag{5.65}$$

In a strong inversion the value of $\lambda(0)$ is very small as compared to T_{Si}, i.e. $\lambda(0) << T_{Si}$. Hence, using $\sin x \approx x$ (for small x) in Equation (5.61), we get

$$\lambda(0) \approx y_0 \tag{5.66}$$

Using Equation (5.65) in Equation (5.63) and solving for $dL_{\psi}/dy_0 = 0$ we get

$$\frac{3}{2}\frac{T_{Si}}{\lambda(0)[T_{Si} + \lambda(0)]} = \frac{\varepsilon_{ox}}{\varepsilon_{Si}T_{ox}} - \frac{4qN_{inv}}{\varepsilon_{Si}\psi_0(0)} \tag{5.67}$$

From Equation (5.67), we can see that $\lambda \sim N_{inv}^{-1}$ when N_{inv} is large and $\lambda(0)$ is small. Hence, $\lambda(0)$ varies faster with N_{inv} in Equation (5.67) than in Equation (5.60).

Hence, using the value of $\lambda(0)$ obtained from Equation (5.67) in Equation (5.48), we get a model for the surface potential of a TFET that incorporates the change in the characteristic length (λ) due to the build-up of inversion charge in the channel, enabling us to capture the change in the potential and the electric fields in the tunnelling region R1 with varying gate bias more accurately.

5.3 The infinite series solution

In this section, we will describe a method for finding the surface potential of a TFET using the separation of variables technique. This method solves the 2D Poisson equation using separation of variables [22], giving us an infinite series solution for the potential. The infinite series obtained is the exact solution of the 2D Poisson equation. However, an infinite series does not give a closed-form solution for the surface potential and hence we take the first term of the infinite series to find the surface potential of the TFET [23–28]. The method is described in detail in the following sections.

5.3.1 Solving the 2D Poisson equation using separation of variables

Let us consider the homogeneous form of the 2D Poisson equation, also known as Laplace's equation:

$$\frac{\partial\psi^2}{\partial x^2} + \frac{\partial\psi^2}{\partial y^2} = 0 \tag{5.68}$$

Let us assume the above equation has the following form of solution:

$$\psi(x, y) = X(x)Y(y) \tag{5.69}$$

where X is a function of x and Y is a function y. Using Equation (5.69) in (5.68) gives

$$X''(x)Y(y) + X(x)Y''(y) = 0$$

which can be written as

$$\frac{X''(x)}{X(x)} = -\frac{Y''(y)}{Y(y)} = \pm\lambda^2 \tag{5.70}$$

where the term $\pm\lambda^2$ is a constant.

The above equation is an eigenvalue problem, with λ as the eigenvalue. For a positive value of the constant term $+\lambda^2$ in Equation (5.70), we get the following set of equations:

$$X''(x) - \lambda^2 X(x) = 0 \tag{5.71}$$

and

$$Y''(y) + \lambda^2 Y(y) = 0 \tag{5.72}$$

Equations (5.71) and (5.72) have the following general form solution:

$$X(x) = C\cosh(\lambda x) + D\sinh(\lambda x) \tag{5.73}$$

and

$$Y(y) = A\cos(\lambda y) + B\sin(\lambda y) \tag{5.74}$$

For a negative value of the constant term $-\lambda^2$ in Equation (5.70), we get the following set of equations:

$$X''(x) + \lambda^2 X(x) = 0 \tag{5.75}$$

and

$$Y''(y) + \lambda^2 Y(y) = 0 \tag{5.76}$$

Equations (5.75) and (5.76) have the following general form solution:

$$X(x) = A\cos(\lambda x) + B\sin(\lambda x) \tag{5.77}$$

and

$$Y(y) = C\cos(\lambda x) + D\sinh(\lambda y) \tag{5.78}$$

Using the solution for $X(x)$ and $Y(y)$ in Equation (5.69), we get the general form solution to the Laplace equation (5.68). However, there are many unknown

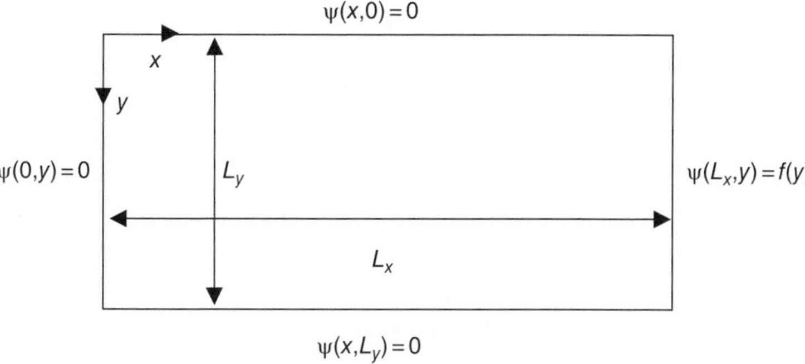

Figure 5.10 Boundary conditions in a 2D solution space for the Laplace equation.

quantities in this solution. The first amongst these is the value of the constant term ($+\lambda^2$or $-\lambda^2$). The sign and the value of λ can be found if we have homogeneous boundary conditions in one direction. Then we have the unknown constants A, B, C and D, which can also be determined by using the boundary conditions. Let us now demonstrate the solution of the 2D Laplace equation for a homogeneous boundary value problem.

5.3.2 Solution of the homogeneous boundary value problem

Let us consider the 2D Laplace equation (5.68) with the following boundary conditions, as shown in Figure 5.10:

$$\psi(x,0)=0 \tag{5.79}$$

$$\psi\left(x,L_y\right)=0 \tag{5.80}$$

$$\psi(0,y)=0 \tag{5.81}$$

$$\psi(L_x,y)=f(y) \tag{5.82}$$

Only one boundary condition in the x direction (5.82) is non-homogeneous and is a function of y; the rest of the boundary conditions are homogeneous.

Since both the boundary conditions in the y direction are homogeneous, choosing $-\lambda^2$ as the constant term of the eigenvalue problem (5.70) gives us a trivial solution for $\psi(x, y)$, as shown below:

$$\psi(x,0)=X(x)Y(0)=0 \Rightarrow Y(0)=0 \tag{5.83}$$

This gives

$$Y(0) = C\cosh(0) + D\sinh(0) = 0 \Rightarrow C = 0 \tag{5.84}$$

Now

$$\psi\left(x, L_y\right) = X(x)Y\left(L_y\right) = 0 \Rightarrow Y\left(L_y\right) = 0 \tag{5.85}$$

which gives

$$Y(0) = D\sinh\left(\lambda L_y\right) = 0 \Rightarrow D = 0 \tag{5.86}$$

Hence

$$Y(y) = 0 \Rightarrow \psi(x, y) = X(x)Y(y) = 0 \tag{5.87}$$

Thus, we get a trivial solution for $\psi(x, y)$.

Hence, we will choose $+\lambda^2$ as the constant term of the eigenvalue problem (5.70), which gives us the general form solution for *X(x)* and *Y(y)* as given in Equations (5.73) and (5.74). Using the boundary conditions in the *y* direction (5.79) and (5.80) gives

$$\psi(x,0) = X(x)Y(0) = 0 \Rightarrow Y(0) = 0 \tag{5.88}$$

$$Y(0) = A\cos(0) + B\sin(0) \Rightarrow A = 0 \tag{5.89}$$

and

$$\psi\left(x, L_y\right) = X(x)Y\left(L_y\right) = 0 \Rightarrow Y\left(L_y\right) = 0 \tag{5.90}$$

$$Y\left(L_y\right) = B\sin\left(\lambda L_y\right) = 0 \Rightarrow \lambda L_y = n\pi;\ n = 1,\ 2,\ 3,\ 4, \ldots \tag{5.91}$$

For each value of n, we will have a different value of B and λ represented by B_n and λ_n, respectively. The value of λ_n is given by

$$\lambda_n = \frac{n\pi}{L_y};\ n = 1,\ 2,\ 3,\ 4, \ldots \tag{5.92}$$

The above equation gives us the eigenvalues λ_n and we get the following expression for $Y(y)$:

$$Y(y) = B_n \sin(\lambda_n y) \tag{5.93}$$

The expression for $\psi(x, y)$ now becomes

$$\psi(x,\ y) = X(x)Y(y) = [C_n \cosh(\lambda_n x) + D_n \sinh(\lambda_n x)]B_n \sin(\lambda_n y) \tag{5.94}$$

We will now apply the boundary condition (5.81), which gives

$$\psi(0,y) = X(0)Y(y) = 0 \Rightarrow X(0) = 0 \tag{5.95}$$

$$X(0) = C_n \cosh(0) + D_n \sinh(0) = 0 \Rightarrow C_n = 0 \tag{5.96}$$

Therefore,

$$X(x) = D_n \sinh(\lambda_n x) \tag{5.97}$$

and

$$X(x)Y(y) = C_n \sinh(\lambda_n x) B_n \sin(\lambda_n y) \tag{5.98}$$

The solution for $\psi(x, y)$ is the linear combination of $X(x)Y(y)$ for all values of λ_n:

$$\psi(x,y) = \sum_{n=0}^{\infty} C_n \sinh(\lambda_n x) B_n \sin(\lambda_n y) \tag{5.99}$$

Combining the coefficients we get:

$$\psi(x,y) = \sum_{n=0}^{\infty} b_n \sinh(\lambda_n x) \sin(\lambda_n y) \tag{5.100}$$

where

$$b_n = C_n B_n \tag{5.101}$$

Now we are left with only one unknown b_n, which can be found using the non-homogeneous boundary condition (5.82):

$$\psi(L_x,y) = \sum_{n=0}^{\infty} b_n \sinh(\lambda_n L_x) \sin(\lambda_n y) \tag{5.102}$$

The above equation is a Fourier series with $b_n \sinh(\lambda_n L_x)$ as the Fourier series coefficient. Hence, b_n can be found by the standard method for finding Fourier coefficients:

$$\int_0^{Ly} \psi(L_x,y) \sin(\lambda_n y) dy = \frac{L_y}{2} b_n \sinh(\lambda_n L_x) \tag{5.103}$$

$$b_n = \frac{2}{L_y \sinh(\lambda_n L_x)} \int_0^{L_y} \psi(L_x, y) \sin(\lambda_n y) dy \tag{5.104}$$

Equation (5.102) can now be written as

$$\psi(x, y) = \sum_{n=0}^{\infty} c_n \frac{\sinh(\lambda_n x)}{\sinh(\lambda_n L_x)} \sin(\lambda_n y) \tag{5.105}$$

Where

$$c_n = \frac{2}{L_y} \int_0^{L_y} \psi(L_x, y) \sin(\lambda_n y) dy \tag{5.106}$$

Equation (5.105) is the solution for the 2D Laplace equation with boundary conditions as given by Equations (5.79) to (5.82). Considering this solution, we will now move on to solving the 2D Poisson equation in a TFET. This will be done by splitting our problem into a linear combination of different boundary value problems, similar to the boundary value problem described in Figure 5.10.

5.3.3 The solution to the 2D Poisson equation in a TFET

Let us now look at the problem of solving the 2D Poisson equation (5.1) in a TFET, which would have the boundary conditions as depicted in Figure 5.11 (a). This problem consists of an inhomogeneous equation (2D Poisson equation) with inhomogeneous boundary conditions. Hence, the method of separation of variables cannot be directly applied to solve this problem. However, since the 2D Poisson equation consists only of linear terms, the problem in hand can be split into a linear combination of simpler problems. Let us now see how this is done.

The problem shown in Figure 5.11(a) can be split into three parts as depicted by Figure 5.9(b) to (d): (i) solving the 1D Poisson's equation in the y-direction satisfying the boundary conditions at $y=0$ and $y=t_{Si}$ (giving $v(y)$), (ii) solving the 2D Laplace equation satisfying the boundary condition at $x=0$ and assuming all other boundary conditions to be homogeneous (giving $u_L(x, y)$) and (iii) solving the 2D Laplace equation satisfying the boundary condition at $x=L_x$ and assuming all other boundary conditions to be homogeneous (giving $u_R(x, y)$).

A linear combination of all these three solutions satisfies the 2D Poisson equation and all the boundary conditions. Hence, we get the following form of solution:

$$\psi(x, y) = v(y) + u_L(x, y) + u_R(x, y) \tag{5.107}$$

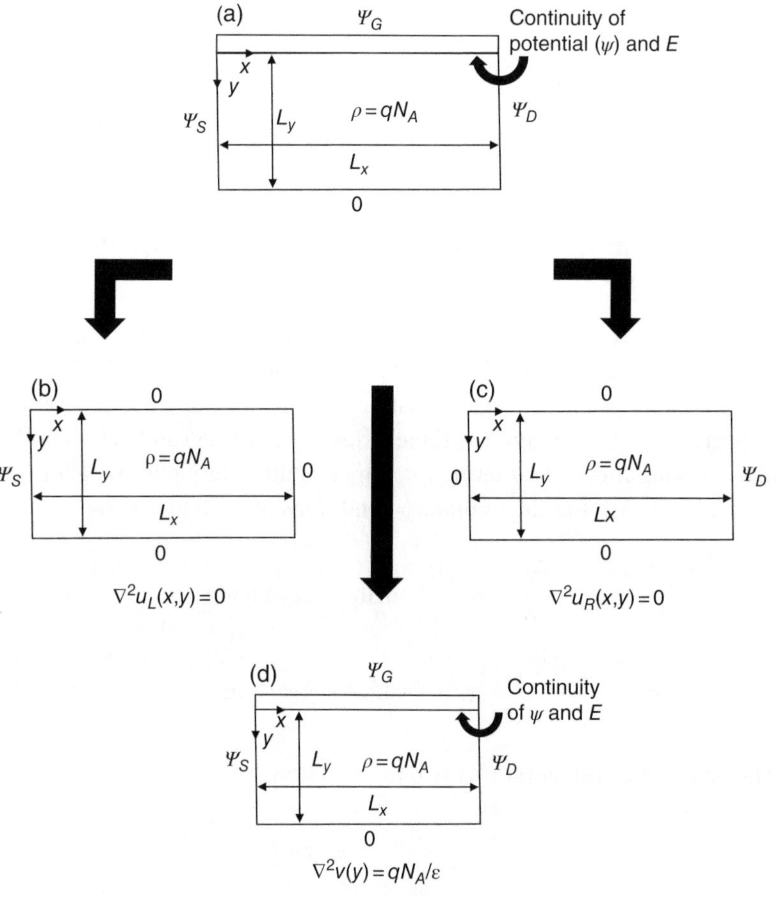

Figure 5.11 Boundary conditions for the potential distribution in a TFET.

Using the above solution in the 2D Poisson equation (5.1), we get

$$\nabla^2\psi(x,y)=\nabla^2v(y)+\nabla^2u_L(x,y)+\nabla^2u_R(x,y)=\frac{qN_A}{\varepsilon_{Si}}+0+0=\frac{qN_A}{\varepsilon_{Si}} \tag{5.108}$$

Hence, the linear combination of the terms $v(y)$, $u_L(x, y)$ and $u_R(x, y)$ satisfies the 2D Poisson equation. This solution (5.107) also satisfies all the boundary conditions of the problem shown in Figure 5.11(a):

$$\psi(x,0)=v(0)+u_L(x,0)+u_R(x,0)=\psi_G-V_{FB}+0+0=\psi_G-V_{FB} \tag{5.109}$$

$$\psi(x,T_{Si})=v(T_{Si})+u_L(x,T_{Si})+u_R(x,T_{Si})=\psi_{Sub}+0+0=\psi_{Sub} \tag{5.110}$$

$$\psi(0,y)=v(y)+u_L(x,0)+u_R(x,0)=v(y)+\psi_{Src}-v(y)+0=\psi_{Src} \tag{5.111}$$

$$\psi(0,y)=v(y)+u_L(x,0)+u_R(x,0)=v(y)+0+\psi_{Drain}-v(y)=\psi_{Drain} \tag{5.112}$$

In the next section, we will solve for the potential in a TFET using the approach outlined here.

5.3.4 The infinite series solution to Poisson's equation in a TFET

Let us now solve for the potential in a TFET using the approach described in the previous section. Figure 5.12 shows the cross-section of an SOI TFET with all its boundary conditions. As given in Equation (5.107), the potential distribution in a TFET is the linear combination of three terms ($v(y)$, $u_L(x, y)$ and $u_R(x, y)$). We will begin by solving for the first term $v(y)$, which is the solution to the 1D Poisson's equation (4.1) satisfying the boundary conditions at $y=0$ and $y=t_{Si}$:

$$\frac{\partial^2 v(y)}{\partial y^2}=\frac{qN_A}{\varepsilon_{Si}} \quad \text{in the silicon body} \tag{5.113}$$

$$\frac{\partial^2 v(y)}{\partial y^2}=0 \quad \text{inside the gate oxid} \tag{5.114}$$

The solution of Equation (5.113) is

$$v(y)=\frac{qN_A}{\varepsilon_{Si}}y^2+\alpha_1 y+\beta_1 \tag{5.115}$$

and the solution of Equation (5.114) is

$$v(y)=\alpha_1 y+\beta_2 \tag{5.116}$$

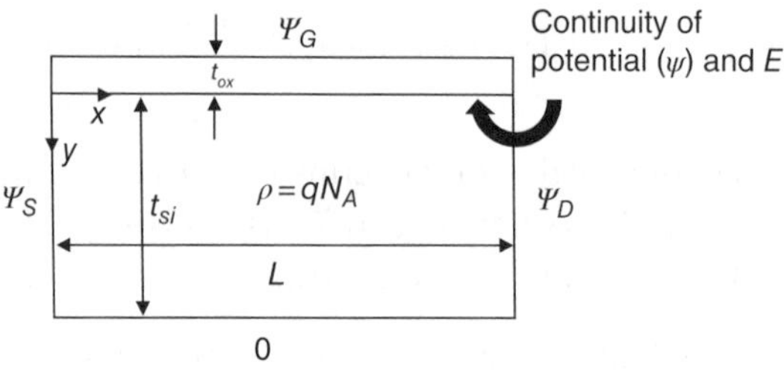

Figure 5.12 Boundary conditions for the potential distribution in an SOI TFET.

Using the boundary conditions shown in (Figure 5.12), we get the following values for α_1, β_1, α_2 and β_2:

$$\alpha_1 = \frac{\varepsilon_{ox}\psi_G + \frac{qN_A\varepsilon_{ox}}{2\varepsilon_{Si}}t_{Si}^2}{\varepsilon_{Si}t_{ox} - \varepsilon_{ox}t_{Si}} \tag{5.117}$$

$$\beta_1 = \beta_2 = \frac{\frac{qN_A}{2}t_{Si}^2 t_{ox} - \psi_G t_{Si}\varepsilon_{ox}}{\varepsilon_{Si}t_{ox} - t_{Si}\varepsilon_{ox}} \tag{5.118}$$

$$\alpha_2 = \frac{\varepsilon_{Si}}{\varepsilon_{ox}}\alpha_1 = \frac{\varepsilon_{Si}\psi_G + \frac{qN_A}{2}t_{Si}^2}{\varepsilon_{Si}t_{ox} - \varepsilon_{ox}t_{Si}} \tag{5.119}$$

Now we will solve for the second term in Equation (5.107), $u_L(x, y)$, which is the solution for the 2D Laplace equation (4.68) with the boundary conditions given in (Figure 5.12). From the separation of variables method described in Section 5.3, we get the following form of solution for the 2D Laplace equation for the boundary value problem described in Figure 5.11(d):

$$\psi(x,y) = \sum_{n=0}^{\infty} c_n \frac{\sinh(\lambda_n x)}{\sinh(\lambda_n L_x)} \sin(\lambda_n y) \tag{5.120}$$

Let us now write a more general form of this equation where the positions of the origin of the sin and sinh functions are also variable:

$$\psi(x,y) = \sum_{n=0}^{\infty} c_n \frac{\sinh(\lambda_n (x - x_i))}{\sinh(\lambda_n L_x)} \sin(\lambda_n (y - y_i)) \tag{5.121}$$

where x_0 and y_0 are the positions of the origin of the sinh and sin functions, respectively.

Here we can observe that the position of y_0 and the value of λ_n satisfy the y-direction boundary conditions and the position of x_0 satisfy the x-direction homogeneous boundary condition. The value of c_n is calculated using the non-homogeneous x-direction boundary condition. We will now find the expression for $u_L(x, y)$ using Equation (5.121) and the boundary conditions given in Figure 5.12. Note that similar to the solution of $v(y)$, we will have different expressions for $u_L(x, y)$ for the silicon body and for the oxide regions. Using the boundary conditions $u_L(x, t_{si}) = 0$ and $u_L(L, y) = 0$, we get the values for y_0 and x_0, respectively. This will give

$$u_L(x,y) = \sum_{n=0}^{\infty} c_n \frac{\sinh(\lambda_n (x - L))}{\sinh(\lambda_n L)} \sin(\lambda_n (y + t_{si})) \quad \text{in the silicon body} \tag{5.122}$$

Similarly, using the boundary conditions for the oxide regions $u_L(x, t_{ox}) = 0$ and $u_L(L, y) = 0$, we get

$$u_L(x,y) = \sum_{n=0}^{\infty} c_{Tn} \frac{\sinh(\lambda_n(x-L))}{\sinh(\lambda_n L)} \sin(\lambda_n(y+t_{ox})) \text{ in the oxide region} \tag{5.123}$$

Now, as given in Figure 5.11(a), $u_L(x, y)$ and $\varepsilon \frac{\partial u_L(x,y)}{\partial y}$ should be continuous at $y = 0$. Using these boundary conditions and equating the nth terms, we get the following set of equations:

$$c_n = \frac{\sinh(\lambda_n(x-L))}{\sinh(\lambda_n L)} \sin(\lambda_n t_{Si}) = -c_{Tn} \frac{\sinh(\lambda_n(x-L))}{\sinh(\lambda_n L)} \sin(\lambda_n t_{ox}) \tag{5.124}$$

$$\lambda_n \varepsilon_{Si} c_n = \frac{\sinh(\lambda_n(x-L))}{\sinh(\lambda_n L)} \cos(\lambda_n t_{Si}) = \lambda_n \varepsilon_{ox} c_{Tn} \frac{\sinh(\lambda_n(x-L))}{\sinh(\lambda_n L)} \cos(\lambda_n t_{ox}) \tag{5.125}$$

Dividing Equation (5.124) by Equation (5.125), we get

$$\frac{\tan(\lambda_n t_{Si})}{\varepsilon_{Si}} = -\frac{\tan(\lambda_n t_{ox})}{\varepsilon_{ox}} \tag{5.126}$$

$$\Rightarrow \varepsilon_{ox} \tan(\lambda_n t_{Si}) + \varepsilon_{Si} \tan(\lambda_n t_{ox}) = 0 \tag{5.127}$$

By solving Equation (5.127), we get the eigenvalues λ_n.

We will now solve for $u_R(x, y)$, which is the solution for the 2D Laplace equation (5.68) with the boundary conditions given in Figure 5.12. Similar to the solution for $u_L(x, y)$ outlined above, we get $u_R(x, y)$ as

$$u_R(x,y) \sum_{n=0}^{\infty} d_n \frac{\sinh(\lambda_n(x))}{\sinh(\lambda_n L)} \sin(\lambda_n(y+t_{si})) \text{ in the silicon region} \tag{5.128}$$

$$u_R(x,y) \sum_{n=0}^{\infty} d_{Tn} \frac{\sinh(\lambda_n(x))}{\sinh(\lambda_n L)} \sin(\lambda_n(y-t_{ox})) \text{ in the oxide region} \tag{5.129}$$

As the y-direction boundary conditions are the same for $u_R(x, y)$ and $u_L(x, y)$, Equations (5.124) and (5.125) will be the same for $u_R(x, y)$ and for $u_L(x, y)$:

$$d_n \frac{\sinh(\lambda_n x)}{\sinh(\lambda_n L)} \sin(\lambda_n t_{Si}) = -d_{Tn} \frac{\sinh(\lambda_n x)}{\sinh(\lambda_n L)} \sin(\lambda_n t_{ox}) \tag{5.130}$$

$$\lambda_n \varepsilon_{Si} d_n \frac{\sinh(\lambda_n x)}{\sinh(\lambda_n L)} \cos(\lambda_n t_{Si}) = \lambda_n \varepsilon_{ox} d_{Tn} \frac{\sinh(\lambda_n x)}{\sinh(\lambda_n L)} \cos(\lambda_n t_{ox}) \tag{5.131}$$

Equations (5.130) and (5.131) give

$$\varepsilon_{ox}\tan(\lambda_n t_{Si})+\varepsilon_{Si}\tan(\lambda_n t_{ox})=0 \tag{5.132}$$

The next task is to find the values of c_n and c_{Tn} in Equations (5.122) and (5.123), respectively. We will first express c_{Tn} in terms of c_n using Equation (5.124):

$$c_{Tn}=-c_n\frac{\sin(\lambda_n t_{Si})}{\sin(\lambda_n t_{ox})} \tag{5.133}$$

Now we can write

$$u_L(x,y)=\sum_{n=0}^{\infty}c_n u_{Ln}(x,\,y) \tag{5.134}$$

Where

$$u_{Ln}(x,y)=\frac{\sinh(\lambda_n(x-L))}{\sinh(\lambda_n L)}\sin(\lambda_n(y+t_{si})) \tag{5.135}$$

$$u_{Ln}(x,y)=-\frac{\sin(\lambda_n t_{Si})}{\sin(\lambda_n t_{ox})}\frac{\sinh(\lambda_n(x-L))}{\sinh(\lambda_n L)}\sin(\lambda_n(y-t_{ox})) \tag{5.136}$$

The value of c_n can now be found using the non-homogeneous x-direction boundary condition and the method for finding Fourier coefficients (as shown in Equation 5.103), as given below:

$$c_n=\frac{\int_{-t_{Si}}^{t_{ox}}[V_{source}-v(y)]g_n(y)dy}{\int_{-t_{Si}}^{t_{ox}}u_{Ln}(0,y)g_n(y)dy} \tag{5.137}$$

where g_n is an orthogonal function of $u_{Ln}(0,\,y)$, such that

$$\int_{-t_{Si}}^{t_{ox}}u_{Ln}(0,y)g_m(y)dy=0;\,n\neq m \tag{5.138}$$

The function g_1 is given by

$$g_1(y) = \sin(\lambda_n(y + t_{si})) \quad \text{in the silicon region} \tag{5.139}$$

$$g_1(y) = \frac{\cos(\lambda_n t_{Si})}{\cos(\lambda_n t_{ox})}\sin(\lambda_n(y - t_{ox})) \quad \text{in the oxide region} \tag{5.140}$$

Hence, we get:

$$c_1 = \frac{\int_{-t_{Si}}^{t_{ox}} [V_{source} - v(y)]g_1(y)dy}{\int_{-t_{Si}}^{t_{ox}} u_{Ln}(0,y)g_1(y)dy} \tag{5.141}$$

Similarly,

$$d_1 = \frac{\int_{-t_{Si}}^{t_{ox}} [V_{drain} - v(y)]g_1(y)dy}{\int_{-t_{Si}}^{t_{ox}} u_{Rn}(0,y)g_1(y)dy} \tag{5.142}$$

By evaluating the above integrals we get the coefficients for the first terms of Equation (5.134). We neglect the higher-order terms (i.e. $n = 2, 3, 4, \ldots$) and get the following expression for $\psi(x, y)$:

$$\psi(x,y) = v(y) + c_1 u_{L1}(x,y) + d_1 u_{R1}(x,y) \tag{5.143}$$

The value of $\psi(x, y)$ at $y = 0$ gives us the surface potential:

$$\psi_S(x) = \psi(x,0) = v(0) + c_1 u_{L1}(x,0) + d_1 u_{R1}(x,0) \tag{5.144}$$

Hence, exploiting the linearity of Poisson's equation and using the method of separation of variables, we get a solution for the surface potential in a TFET. This method has been extensively used in various analytical models reported to date for the drain current of a TFET. This method, however, assumes the entire channel to be depleted, which may not be the case with long channel devices. In such a case, we can solve for the 2D Poisson equation only in the tunnelling region with the appropriate boundary conditions.

5.4 Extension of surface potential models to different TFET structures

All the methods for finding the surface potential discussed so far have been derived for a single gate SOI TFET. In this section, we will discuss extending these models to double gate (DG) and gate all around (GAA) TFET structures. The advantages of these structures have been discussed previously in Section 3.3.

5.4.1 DG TFET

Let us consider a double gate (DG) TFET structure as shown in Figure 5.13. The device has a channel length of L, silicon film thickness of T_{Si}, oxide thickness T_{ox} and a body doping of N_A. The device has a long channel (i.e. channel length > 50 nm) and a thin silicon film ($T_{Si} \sim 10$ nm) such that the body is fully depleted under the influence of a gate voltage. The top and the bottom gates are of the same material and hence the structure is symmetric along the a–a' axis. This device has two channels, one at the top silicon–oxide interface and other at the bottom silicon–oxide interface. Since the device is symmetric, we will only study one of the two channels and use the same results for the other channel. However, the reader should note that the electrostatics of a single gate and a double gate TFET are not the same. Hence, the model results derived for a single gate TFET in the previous sections cannot be directly applied to a DG TFET.

Let us first discuss the pseudo-2D method for finding the surface potential in a DG TFET [15, 17]. We start by using the parabolic potential approximation (5.3) in a DG TFET. Figure 5.14 shows the potential along the y direction in a DG TFET. From this figure, we can observe that the potential distribution in the y-direction in a DG TFET has the following boundary conditions:

(i) The electric field at the centre of the body (i.e. $y = 0$) is equal to zero, which gives

$$\left.\frac{\partial \psi(x,y)}{\partial y}\right|_{y=0} = 0 \tag{5.145}$$

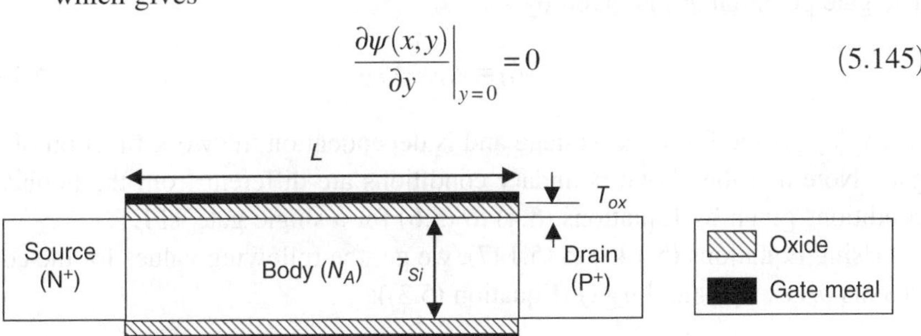

Figure 5.13 Schematic of a p-channel double gate (DG) TFET.

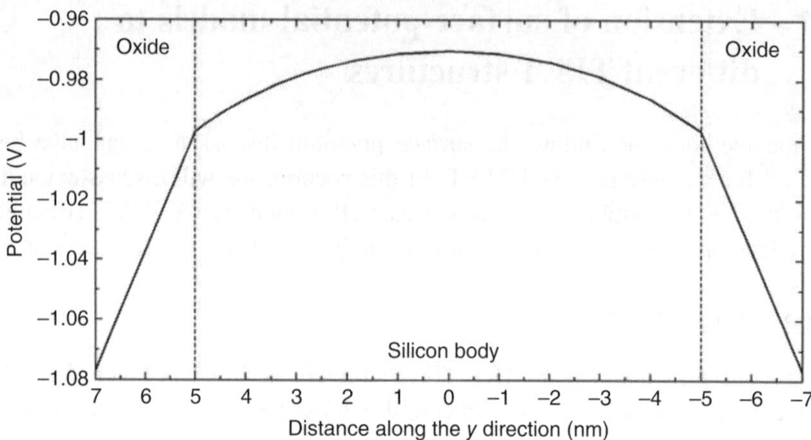

Figure 5.14 Potential distribution along the y-direction in a DG TFET.

(ii) The potential at the Si–SiO_2 interface (i.e. $y = t_{Si}/2$) is equal to the surface potential ($\psi_S(x)$), which gives

$$\psi(x, t_{si}/2) = \psi_S(x) \tag{5.146}$$

(iii) The electric field displacement is continuous across the Si–SiO_2 interface (i.e. $y = t_{Si}/2$), which gives

$$\left.\frac{\partial\psi(x,y)}{\partial y}\right|_{y=t_{SI}/2} = \frac{C_{ox}(\psi_G - \psi_S(x))}{\varepsilon_{Si}} \tag{5.147}$$

where C_{ox} is the gate oxide capacitance per unit area and ψ_G is the gate potential. The gate potential ψ_G is given by

$$\psi_G = V_{GS} - V_{FB} \tag{5.148}$$

where V_{FB} is the flat-band voltage and is dependent on the work function of the gate. Note that the above boundary conditions are different from the boundary conditions given by Equations (5.4) to (5.6) for a single gate TFET.

Using Equations (5.145) to (5.147), we get the following values for the coefficients $a_0(x)$, $a_1(x)$ and $a_2(x)$ (Equation (5.3)):

$$a_0(x) = \psi_S(x)\left(1 + \frac{C_{ox}t_{Si}}{4\varepsilon_{Si}}\right) - \frac{\psi_G t_{ox}}{4t_{Si}} \tag{5.149}$$

$$a_1(x) = 0 \tag{5.150}$$

$$a_2(x) = \frac{C_{ox}(\psi_G - \psi_S(x))}{t_{Si}\varepsilon_{Si}} \tag{5.151}$$

Using the above coefficients in the 2D Poisson equation (5.1), gives us the following second-order differential equation in the surface potential:

$$\frac{\partial^2 \psi_S(x)}{\partial x^2} - \frac{2C_{ox}}{t_{Si}\varepsilon_{Si}}\psi_S(x) = \frac{qN_A}{\varepsilon_{Si}} - \frac{2C_{ox}}{t_{Si}\varepsilon_{Si}}\psi_G \tag{5.152}$$

This equation is of the same form as Equation (5.14) derived in Section 5.1.2 for a single gate TFET, but with a difference in the coefficients of $\psi_S(x)$ and ψ_G terms. Hence, the final solution for the surface potential of a DG TFET will have the same form as that of a single gate TFET as given by Equations (5.33) and (5.34) with a change in the value of the parameter L_d (i.e. the characteristic length), which is

$$L_d = \sqrt{t_{Si}\varepsilon_{Si}/2C_{ox}} = \sqrt{t_{Si}t_{ox}\varepsilon_{Si}/2\varepsilon_{ox}} \text{ for a DG TFET} \tag{5.153}$$

The increase in the drain in a DG TFET as compared to a single gate TFET is due to two factors, the first due to the fact that we have two parallel channels in a DG TFET and the second due to a decrease in the value of L_d for a double gate TFET as compared to a single gate TFET. Hence, the drain current of a DG TFET will be more than twice of that of a single gate TFET, with the same device parameters.

Let us now recall the other approaches for finding the surface potential discussed in the previous sections.

The variational approach discussed in Section 5.2 can be extended for finding the surface potential of a DG TFET by using the symmetry of the device along the *a–a′* axis and transforming the problem into a single TFET with appropriate integration limits, as depicted in Figure 5.15. As outlined in Section 5.2, the use of this

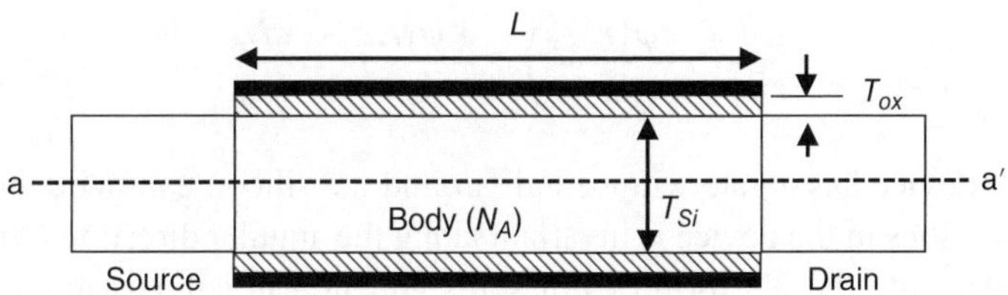

Figure 5.15 Schematic of a p-channel double gate (DG) TFET showing the axis of symmetry.

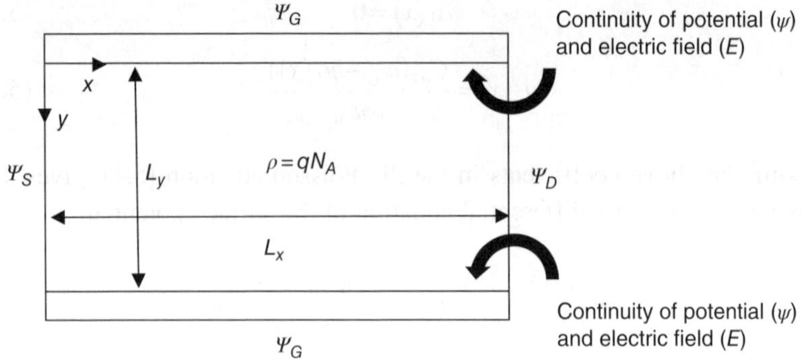

Figure 5.16 Boundary conditions for the potential distribution of a DG TFET.

method helps in incorporating the effect of inversion charge in the surface potential of the TFET.

The infinite series method discussed in Section 5.3 can also be extended to a DG TFET. In this case, we need to follow the same procedure as outlined in Section 5.3 on the boundary value problem depicted in Figure 5.16. As discussed in Section 5.3, the infinite series method is the exact solution to the 2D Poisson equation, giving us a more complex but accurate model for the surface potential of the TFET.

We will now discuss how to find the surface potential for a gate all around (GAA) nanowire TFET.

5.4.2 GAA TFET

A GAA TFET as described in Section 3.3.2 is a three-dimensional structure with a gate wrapped all around a piece of silicon nanowire as shown in Figure 5.17. Since, the device has a cylindrical structure, we need to solve for the 2D Poisson equation in cylindrical coordinates, as given below:

$$\frac{1}{r}\frac{\partial}{\partial r}\left(r\frac{\partial\psi(r,z)}{\partial r}\right)+\frac{\partial^2\psi(r,z)}{\partial z^2}=\frac{qN_A}{\varepsilon_{Si}} \tag{5.154}$$

As the device has a gate wrapped all around the silicon nanowire uniformly, the electrostatics in the device is invariant along the angular direction (ϕ). Hence, we consider only the 2D form of Poisson's equation in cylindrical coordinates (5.154), neglecting the angular dimension. If we try to draw an analogy between a planar and a cylindrical structure then the r-direction corresponds to the y-direction, the z-direction corresponds to the x-direction and the angular direction corresponds to the z-direction.

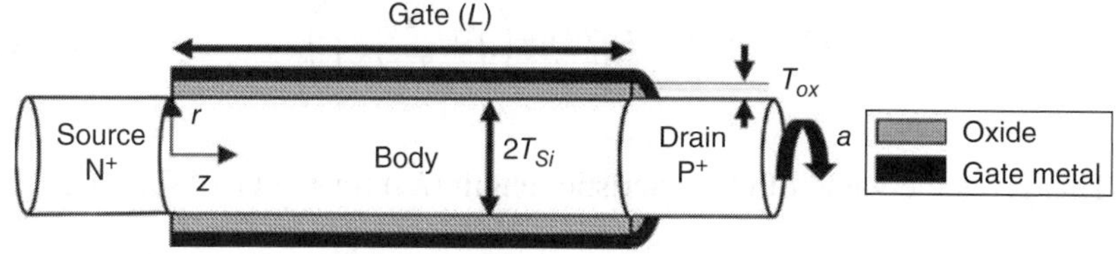

Figure 5.17 Schematic of a p-channel gate all around (GAA) nanowire TFET.

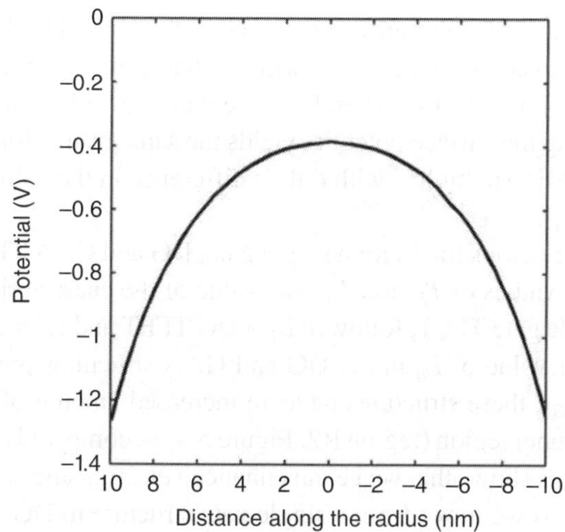

Figure 5.18 Potential distribution along the radius of a GAA nanowire TFET.

Now let us look at the potential distribution of a GAA nanowire TFET in the *r*-direction (Figure 5.18). The shape of the potential distribution of a GAA nanowire TFET in the *r*-direction is exactly the same as that of a DG TEFT in the *y*-direction. Hence, the *r*-direction boundary conditions for a GAA nanowire TFET are exactly the same as those given by Equations (5.145) to (5.147). Hence, using the pseudo-2D method in a GAA nanowire TFET gives us the following second-order differential equation in the surface potential:

$$\frac{\partial^2 \psi_S(z)}{\partial z^2} - \frac{2C_{ox}}{T_{Si}\varepsilon_{Si}}\psi_S(z) = \frac{qN_A}{\varepsilon_{Si}} - \frac{2C_{ox}}{T_{Si}\varepsilon_{Si}}\psi_G \tag{5.155}$$

Equation (5.155) is exactly the same as Equation (5.152), with only a change in the coordinates. However, for a GAA nanowire TFET the expression for C_{ox} will be different from that of a DG TFET. The parameter C_{ox} for a GAA nanowire TFET is the gate capacitance per unit area at the inner surface of the oxide (i.e. at $r = T_{Si}/2$) and is given by

$$C_{ox} = \varepsilon_{ox} / \left((T_{Si}/2) \ln \left(1 + \frac{T_{ox}}{T_{Si}/2} \right) \right) \tag{5.156}$$

This gives the value of characteristic length (L_d) for the GAA TFET as

$$L_d = \sqrt{T_{Si}^2 \ln \left(1 + \frac{2T_{ox}}{T_{Si}} \right) \varepsilon_{Si} / (4\varepsilon_{ox})} \tag{5.157}$$

Hence the solution for the surface potential of the GAA TFET will also be the same as that outlined in Section 5.1 with a characteristic length as given by Equation (5.157) [10, 11, 14]. Therefore, we can conclude that the pseudo-2D method for finding the surface potential yields the same results for the single gate, DG and GAA TFET structures, with only a difference in the value of the characteristic length (L_d).

From the expressions for L_d for a single gate, DG and GAA TFET, we can see that for the same values of T_{Si} and T_{ox}, the value of the characteristic length L_d is largest for a single gate TFET, followed by a DG TFET and is smallest for a GAA TFET. A smaller value of L_d in the DG and GAA structures means that for the same value of V_{GS}, these structures have an increased amount of inversion layer charge in the channel region (region R2, Figure 5.4) as compared to that in a single gate structure. This shows that we get an enhanced electrostatic control of the gate over the channel as we move from a single gate structure to DG and GAA structures. Thus, DG and GAA structures exhibit lesser short channel effects (SCEs) as compared to a single gate device. In a TFET, a smaller value of L_d also means that there will be a higher x-direction electric field in the tunnelling region in these device structures leading to an enhanced drain current.

We will now briefly outline how the other approaches discussed in the previous sections can be applied for finding the surface potential of a GAA TFET.

The variational approach discussed in Section 5.2 can be used in a GAA nanowire TFET, by first formulating the variational form of the 2D cylindrical Poisson's equation and then by following a procedure similar to that outlined in Section 5.2. The use of this method helps in incorporating the effect of inversion charge in the surface potential of the GAA TFET.

For using the infinite series method to find the surface potential in a GAA nanowire TFET, one has to first find the general form solution for the 2D Poisson equation in cylindrical coordinates using the method of separation of variables and then use a boundary value problem as outlined for the DG TFET in the previous section. As discussed in Section 5.3, the infinite series method is the exact solution to the 2D Poisson's equation, giving us a more complex but accurate model for the surface potential of the TFET.

We will now discuss the surface potential for a dual material gate (DMG) TFET, using the pseudo-2D method.

5.4.3 Dual material gate TFET

5.4.3.1 Pseudo-2D model for a dual material gate TFET

As discussed in Section 3.3.1, a dual material gate (DMG) TFET is an important TFET architecture and provides many potential advantages. In this section, we will discuss about modelling the surface potential in a DMG TFET using the pseudo-2D model [9, 10]. Figure 5.19 shows the structure of a DMG TFET and also a simulated surface potential curve. As can be seen in the figure, the DMG TFET has three depletion regions in the body, the source–channel depletion region (region RD1) and two depletion regions at the interface of the two gates,

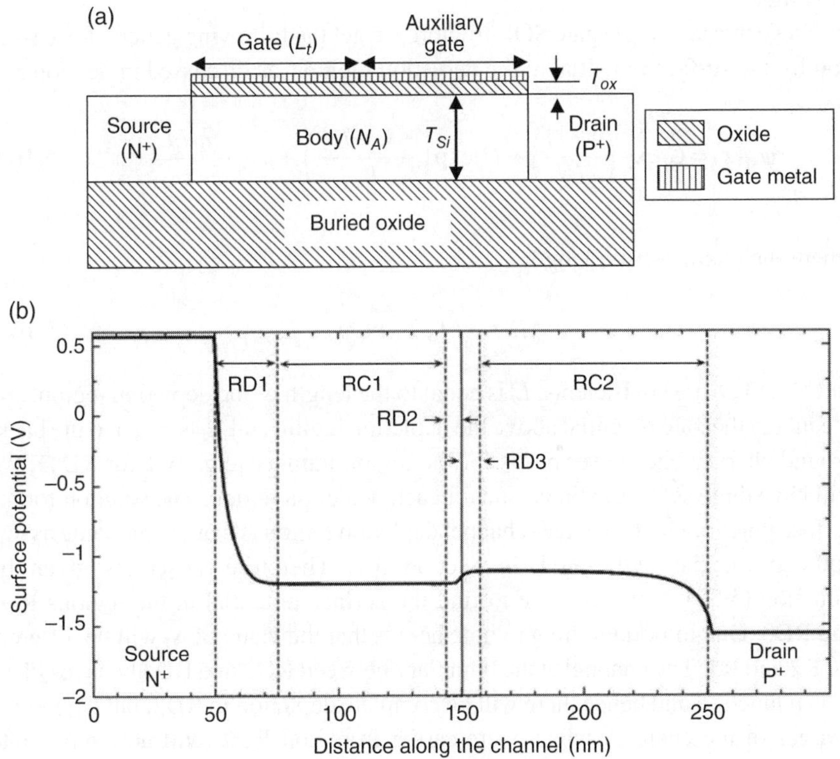

Figure 5.19 (a) Schematic of a p-channel dual material gate (DMG) TFET. (b) Surface potential along the channel of a DMG TFET.

one under the tunnelling gate (RD2) and one under the auxiliary gate (RD3). Also there are two channel regions, one under the tunnelling gate (RC1) and one under the auxiliary gate (RC2). Similar to Equations (5.20 and 5.21), the surface potential in regions RC1 and RC2 can be written as:

$$\psi_{Ct} = V_D + \psi_{Bt} \text{ in the saturation region} \tag{5.158}$$

$$\psi_{Ct} = V_{GS} - V_{tht} + \psi_{Bt} \text{ in the linear region} \tag{5.159}$$

$$\psi_{Ca} = V_D + \psi_{Ba} \text{ in the saturation region} \tag{5.160}$$

$$\psi_{Ca} = V_{GS} - V_{tha} + \psi_{Ba} \text{ in the linear region} \tag{5.161}$$

where in the symbol subscripts t is for the region under the tunnelling gate (RC1) and a is for the region under the auxiliary gate (RC2); ψ_B is the built-in potential of the channel, which is the sum of the band bending in the body and drop across the buried oxide, and V_{th} is the threshold voltage of a MOSFET of an equivalent structure.

Since this is a single gate SOI structure we get the following general form solution for the surface potential in the depletion regions, as discussed in Section 5.1:

$$\psi_{si}(x) = C_i \exp\left(\frac{x-L_i}{L_d}\right) + D_i \exp\left(\frac{-(x-L_i)}{L_d}\right) + \psi_{Gi} - \frac{qN_iL_d^2}{\varepsilon_{Si}} \tag{5.162}$$

where the characteristic length L_d is

$$L_d = \sqrt{T_{Si}T_{ox}\varepsilon_{Si}/\varepsilon_{ox}} \tag{5.163}$$

and C_i and D_i are coefficients, L_i is equal to the length of the depletion region, ψ_{Gi} is equal to the gate potential above the depletion region and N_i is equal to the background charge. The subscript i is for the region number (e.g. $i = 1$ for RD1). We will have three such equations, one for each depletion region. The solution for the surface potential in the source–channel depletion region is exactly the same as that outlined for the SOI TFET in Section 5.1. Therefore, $\psi_{s1}(x)$ is given by Equation (5.33). Now we have to find the surface potential in the regions RD2 and RD3. One important thing to note here is that the value of N_i will be different for R2 and R3. The channel at the boundary between RD2 and RD3 behaves like a p^+-p junction and hence there will be complete depletion in RD2, but the mobile charges of the channel, which were earlier in region RD2, will now move into region RD3. We assume that the density of this charge in region RD3 to be n cm^{-3}. Hence, N_i in RD2 will be the body doping N_A and in RD3 it will be $N_A + n$. We will first define the position of $x = 0$ at the interface of the two gates.

Note that this has been done to get simpler expressions for the coefficients of Equation (5.162). We will then use the following boundary conditions for finding the unknowns C_2, D_2, C_3, D_3, L_2 and L_3.

(i) The value of ψ_S at $x = -L_2$ is equal to ψ_{Ct} and at $x = L_3$ is equal to ψ_{Ca}:

$$\psi_{s2}(-L_2) = \psi_{Ct} \tag{5.164}$$

$$\psi_{s3}(-L_3) = \psi_{Ca} \tag{5.165}$$

(ii) The electric field at $x = -L_2$ and at $x = L_3$ is zero:

$$\frac{\partial \psi_{s2}(-L_2)}{\partial z} = 0 \tag{5.166}$$

$$\frac{\partial \psi_{s3}(L_3)}{\partial z} = 0 \tag{5.167}$$

(iii) The surface potential is continuous at $x = 0$:

$$\psi_{s2}(0) = \psi_{s3}(0) \tag{5.168}$$

(iv) The electric field is continuous at $x = 0$:

$$\frac{\partial \psi_{s2}(0)}{\partial z} = \frac{\partial \psi_{s3}(0)}{\partial z} \tag{5.169}$$

From Equations (5.164) and (5.166), we get

$$C_2 = D_2 = \frac{\psi_{Ct} - \psi_{Gt} + \frac{qN_S L_d^2}{\varepsilon_{Si}}}{2} \tag{5.170}$$

From Equations (5.165) and (5.167), we get

$$C_3 = D_3 = \frac{\psi_{Ca} - \psi_{Ga} + \frac{q(N_S + n)L_d^2}{\varepsilon_{Si}}}{2} \tag{5.171}$$

From Equation (5.168), we get

$$\begin{aligned} &\frac{\psi_{Ct} - \psi_{Gt} + \frac{qN_S L_d^2}{\varepsilon_{Si}}}{2}\left(e^{L_2/L_d} + e^{-L_2/L_d}\right) + \psi_{Gt} - \frac{qN_S L_d^2}{\varepsilon_{Si}} \\ &= \frac{\psi_{Ca} - \psi_{Ga} + \frac{q(N_S + n)L_d^2}{\varepsilon_{Si}}}{2}\left(e^{L_3/L_d} + e^{-L_3/L_d}\right) + \psi_{Ga} - \frac{q(N_S + n)L_d^2}{\varepsilon_{Si}} \end{aligned} \tag{5.172}$$

From Equation (5.169), we get

$$\frac{\psi_{Ct}-\psi_{Gt}+\dfrac{qN_S L_d^2}{\varepsilon_{Si}}}{2}\left(e^{L_2/L_d}-e^{-L_2/L_d}\right)=\frac{\psi_{Ca}-\psi_{Ga}+\dfrac{q(N_S+n)L_d^2}{\varepsilon_{Si}}}{2}\left(e^{-L_3/L_d}-e^{L_3/L_d}\right) \tag{5.173}$$

The value of n in region R3 can be calculated by using the following condition, which comes from the conservation of charge across the p^+-p junction at the boundary of the tunnelling gate and the auxiliary gate, that is

$$nL_3 = n_{ch2}L_2 \tag{5.174}$$

where n_{ch2} is the channel inversion charge concentration under the tunnelling gate. Simultaneously solving (5.172), (5.173) and (5.174) gives us the values of L_2, L_3 and n. These equations have to be solved numerically as they contain non-linear expressions. Since, after the onset of strong inversion, the channel charge does not vary greatly and L_2 and L_3 are of the same order, we can simplify our model by assuming n to be constant. As the inversion charge in strong inversion mostly remains a constant with the applied V_{GS}, we can assume n to be fixed at $10^{19}/\text{cm}^3$ as n_{ch2} is typically $10^{19}/\text{cm}^3$ in strong inversion. Now we only need to solve Equations (5.172) and (5.173) and find L_2 and L_3.

We can now write the final expression for the surface potential of the DMG TFET, as follows:

$$\psi_s(x)=(\psi_{Ct}-\psi_{Gt})\cosh\left(\frac{x-L_d\cosh^{-1}(V_{bi}/(\psi_{Gt}))}{L_d}\right)+\psi_G-\frac{qN_A L_d^2}{\varepsilon_{Si}} \quad \text{in region RD1} \tag{5.175}$$

$$\psi_s=\psi_{Ct}, \quad \text{in region RC1} \tag{5.176}$$

$$\psi_s(x)=C_2\exp\left(\frac{x-L_2}{L_d}\right)+D_2\exp\left(\frac{-(x-L_2)}{L_d}\right)+\psi_{Gt}-\frac{qN_A L_d^2}{\varepsilon_{Si}} \quad \text{in region RD2} \tag{5.177}$$

$$\psi_s(x)=C_3\exp\left(\frac{x-L_3}{L_d}\right)+D_3\exp\left(\frac{-(x-L_3)}{L_d}\right)+\psi_{Ga}-\frac{q(N_A+n)L_d^2}{\varepsilon_{Si}} \quad \text{in region RD3} \tag{5.178}$$

$$\psi_s=\psi_{Ca} \quad \text{in region RC2} \tag{5.179}$$

Most of the studies on DMG TFETs to date suggest that L_t should be much smaller than L_a. Therefore, let us extend our model for a more general case where L_t is much smaller than L_a. In such a case, the entire length of the channel under the tunnelling gate may be depleted, i.e. regions RD1 and RD2 merge into each other and region RC1 does not exist. This will happen more so at low values of gate voltages where L_1 and L_2 are larger. In a structure where L_t is small, typically below 20 nm, we will first solve Equations (5.164) to (5.174) and check if $L_1 + L_2$ is larger than L_t. If it is true then we modify our surface potential models. Now we have only two depletion regions: RD1, which is the entire region under the tunnelling gate, and RD3, which is the same as earlier. We will now have two equations like (5.162) and six unknowns (C_1, D_1, L_1, C_3, D_3 and L_3). We will again use six boundary conditions as earlier defining $x = 0$ as the junction of the two gates.

The tunnelling gate length L_t is the length of region RD1 now and not L_1. Since regions RD1 and RD2 have merged, we will get a point of minimum in the surface potential at $x = -L_1$. As a result, the condition given by (5.164) will be different now and will be as follows:

$$\psi_{S1}(-L_t) = V_{bi} \tag{5.180}$$

The other five boundary conditions given by Equations (5.165) to (5.169) remain the same but the variables and constants of region RD2 are replaced by those of region RD1 (i.e. ψ_{s2} will become ψ_{s1} and so on). Solving as done earlier, we get the following:

$$C_3 = D_3 = \frac{\psi_{Ca} - \psi_{Ga} + \frac{q(N_S + n)L_d^2}{\varepsilon_{Si}}}{2} \tag{5.181}$$

$$C_1 = D_1 = \left(V_{bi} - \psi_{Gt} + \frac{qN_S L_d^2}{\varepsilon_{Si}}\right) \Big/ \left(2\cosh\left(\frac{-L_t + L_1}{L_d}\right)\right) \tag{5.182}$$

$$\begin{aligned}&\frac{V_{bi} - \psi_{Gt} + \frac{qN_S L_d^2}{\varepsilon_{Si}}}{2\cosh\left(\frac{-L_t + L_1}{L_d}\right)}\left(e^{L_1/L_d} + e^{-L_1/L_d}\right) + \psi_{Gt} - \frac{qN_S L_d^2}{\varepsilon_{Si}} \\ &= \frac{\psi_{Ca} - \psi_{Gt} + \frac{q(N_S + n)L_d^2}{\varepsilon_{Si}}}{2}\left(e^{L_3/L_d} + e^{-L_3/L_d}\right) + \psi_{Ga} - \frac{q(N_S + n)L_d^2}{\varepsilon_{Si}}\end{aligned} \tag{5.183}$$

$$\frac{V_{bi} - \psi_{Gt} + \frac{qN_S L_d^2}{\varepsilon_{Si}}}{2\cosh\left(\frac{-L_t + L_1}{L_d}\right)}\left(e^{L_1/L_d} - e^{-L_1/L_d}\right) = \frac{\psi_{Ca} - \psi_{Ga} + \frac{q(N_S + n)L_d^2}{\varepsilon_{Si}}}{2}\left(e^{-L_3/L_d} - e^{L_3/L_d}\right) \tag{5.184}$$

Simultaneously solving Equations (5.182) to (5.184) gives us L_1, L_3 and C_1.

We can now write the final expression for the surface potential of the DMG TFET, in this case:

$$\psi_s(x) = C_1 \exp\left(\frac{x-L_1}{L_d}\right) + D_1 \exp\left(\frac{-(x-L_1)}{L_d}\right) + \psi_{Gt} - \frac{qN_A L_d^2}{\varepsilon_{Si}} \quad \text{in region RD1} \tag{5.185}$$

$$\psi_s(x) = C_3 \exp\left(\frac{x-L_3}{L_d}\right) + D_3 \exp\left(\frac{-(x-L_3)}{L_d}\right) + \psi_{Ga} - \frac{q(N_A+n)L_d^2}{\varepsilon_{Si}} \quad \text{in region RD3} \tag{5.186}$$

$$\psi_s = \psi_{Ca} \quad \text{in region RC2} \tag{5.187}$$

We will now outline the infinite series method for finding the surface potential in a DMG TFET.

5.4.3.2 Infinite series method for DMG TFET

The infinite series method discussed in Section 5.3 can also be used for modelling the surface potential of a DMG TFET [24, 25]. In this section, we will provide a brief outline for using the infinite series method for modelling the surface potential of a DMG TFET. For this purpose, we will divide the structure of the DMG TFET into two parts along the boundary between the two gates and use the infinite series method individually in each part. This will give us two equations like Equation (5.107), as given below:

$$\psi_t(x,y) = v_1(y) + u_{L1}(x,y) + u_{R1}(x,y) \tag{5.188}$$

$$\psi_a(x,y) = v_2(y) + u_{L2}(x,y) + u_{R2}(x,y) \tag{5.189}$$

where $\psi_t(x, y)$ and $\psi_a(x, y)$ are the potential distributions under the tunnelling gate and the auxiliary gate, respectively.

We will first solve for the 1D Poisson equation in each of the two structures shown in Figure 5.20. This gives us $v_1(y)$ and $v_2(y)$, which are the solution to the 1D Poisson equation. We then solve for the 2D Laplace equation in each of the two structures to get $u_{L1}(x, y)$, $u_{R1}(x, y)$, $u_{L2}(x, y)$ and $u_{R2}(x, y)$. Their general form solutions are:

$$u_{L1}(x,y) \sum_{n=0}^{\infty} c_{n1} \frac{\sinh(\lambda_n(x-L_1))}{\sinh(\lambda_n L_1)} \sin(\lambda_n(y+t_{Si})) \tag{5.190}$$

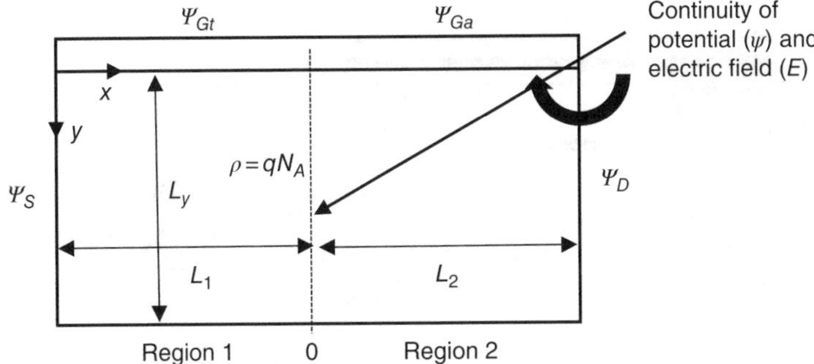

Figure 5.20 Boundary conditions for the potential distribution of a DMG TFET.

$$u_{R1}(x,y)\sum_{n=0}^{\infty} d_{n1}\frac{\sinh(\lambda_n x)}{\sinh(\lambda_n L_1)}\sin(\lambda_n(y+t_{Si})) \tag{5.191}$$

$$u_{L2}(x,y)\sum_{n=0}^{\infty} c_{n2}\frac{\sinh(\lambda_n(x-(L_1+L_2)))}{\sinh(\lambda_n(L_1+L_2))}\sin(\lambda_n(y+t_{Si})) \tag{5.192}$$

$$u_{R2}(x,y)\sum_{n=0}^{\infty} d_{n2}\frac{\sinh(\lambda_n(x-L_1))}{\sinh(\lambda_n(L_1+L_2))}\sin(\lambda_n(y+t_{Si})) \tag{5.193}$$

The values of the coefficients c_{n1}, d_{n1}, c_{n2} and d_{n2} can be found using source and drain side boundary conditions and the continuity of electric field and potential at the interface between the two gates. Exact working out of this model is beyond the scope of this book but readers are encouraged to further explore this model.

We will now move towards modelling the surface potential in a TFET in the presence of localised charges at the silicon–oxide interface.

5.5 The effect of localised charges on the surface potential

TFETs are highly prone to hot-carrier effects at the source–channel region due to high lateral electric fields. These hot carriers get injected into the gate creating permanent damage in the oxide, leading to the presence of localised charges at the silicon–oxide interface. This changes the flat-band voltage of the gate above the damaged region, which causes a change in the surface potential profile in the tunnelling of a TFET and a change in the drain current [12]. Hence, it is important

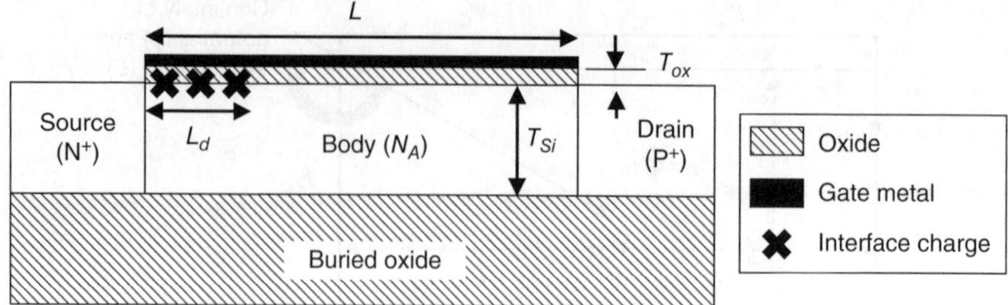

Figure 5.21 Schematic of a p-channel dual soi TFET showing the presence of localised charges at the si-sio₂ interface near the source side.

to model the surface potential in the presence of localised charges at the silicon–oxide interface near the source side of a TFET, as shown in Figure 5.21. The approach followed to model the surface potential in a DMG TFET can also be extended here [16, 19]. Let us assume the density of the localised charges to be N_f and the length of the affected region to be L_D. The flat-band voltage of the gate in the presence of localised charges is given by

$$V_{FB} = V_{FB0} + \frac{qN_f}{C_{ox}} \tag{5.194}$$

where V_{FB0} is the flat-band voltage of the gate without any localised charge. Hence, the surface potential in a TFET in the presence of localised charges can be modelled as the surface potential in a DMG TFET having two gates with the following gate potentials:

$$\psi_{G1} = V_{GS} - V_{FB} = V_{GS} - V_{FB0} + \frac{qN_f}{C_{ox}} \quad \text{equivalent to the tunnelling gate} \tag{5.195}$$

and

$$\psi_{G2} = V_{GS} - V_{FB} = V_{GS} - V_{FB0} \quad \text{equivalent to the auxiliary gate} \tag{5.196}$$

5.6 Surface potential in the depletion regions

In the surface potential model described so far, we have neglected the depletion region in the source side at the source–body interface. We assumed the surface potential to be constant (V_{bi}) throughout the source region. Such an assumption

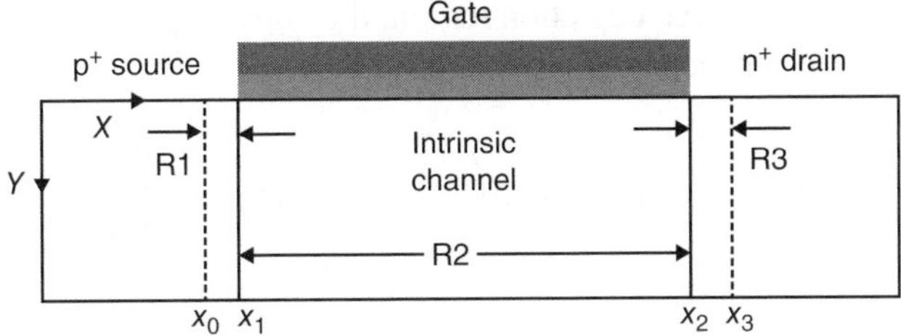

Figure 5.22 Schematic of an n-channel TFET showing source and drain depletion regions.

holds very well for high source doping (which is generally the case with TFETs). However, as a more general case, let us now discuss modelling the surface potential inside the source depletion region.

We will now calculate the surface potential in the depletion regions in the device and observe the effects that inclusion of the depletion regions has on the surface potential within the channel.

Figure 5.22 shows the schematic of the device that is being modelled. The device is divided into three regions – the source side depletion region (R1), the channel (R2) and the drain side depletion region (R3). Due to a different doping in each region, Poisson's equation is written separately in each of the three regions:

$$\frac{\partial^2 \psi_{s,i}}{\partial x^2} + \frac{\partial^2 \psi_{s,i}}{\partial y^2} = \frac{-qN_i}{\epsilon_{Si}} \tag{5.197}$$

where $\psi_{s,i}$ and N_i are the surface potential and doping in the ith region, respectively. Using the pseudo-2D method outlined in the preceding section, the surface potential is approximated as a parabola in the y-direction in each region. The boundary conditions used (Equations (5.4) and (5.5)) to determine the coefficients of this parabola will also be the same as described in the preceding section. However, the boundary condition relating to the continuity of the vertical electric displacement (Equation (5.6)) will be modified in the depletion regions. This is because the gate-body capacitance is different for the channel (C_{ox}) and for the depletion regions. Therefore, in the depletion regions, we must use the fringing field capacitance ($C_{ox,f} = (2/\pi)C_{ox}$) of the gate [6, 29]. For example, in the source depletion region (R1):

$$\left.\frac{\partial \psi_{s,1}}{\partial y}\right|_{y=0} = \frac{-C_{ox,f}\left(\psi_G - \psi_{s,1}\right)}{\epsilon_{Si}} \tag{5.198}$$

Therefore, in the source depletion region, the above equation leads to the following expression for the surface potential:

$$\psi_{s,1} = C_1 \exp\left(\frac{x-x_0}{L_{d,1}}\right) + D_1 \exp\left(-\frac{x-x_0}{L_{d,1}}\right) + \psi_G - \frac{2qN_1 L_{d,1}^2}{\pi\epsilon_{si}} \tag{5.199}$$

$$L_{d,1} = \sqrt{\frac{\pi t_{Si} t_{ox} \epsilon_{Si}}{2\epsilon_{ox}}} \tag{5.200}$$

Similar equations can be written in the drain depletion region, giving $\psi_{s,3}$ in the drain depletion region (R3). The surface potential in the channel $\psi_{s,2}$ will have the same form as previously discussed (Equation (5.26). Now that we have the surface potentials in all the three regions, we need to solve for the coefficients C_i and D_i and depletion region lengths L_1 and L_3 by writing the electrostatic boundary conditions in the horizontal direction. Beginning from the source side, the potential should be equal to the source potential at $x = x_0$ and the field at the depletion region edge should be zero:

$$\psi_{s,1}(x_0) = V_S + V_{bi,1} \tag{5.201}$$

$$\left.\frac{\partial \psi_{s,1}}{\partial x}\right|_{x=x_0} = 0 \tag{5.202}$$

Similar boundary conditions can be written on the drain side depletion region, at $x = x_3$:

$$\psi_{s,3}(x_3) = V_{DS} + V_{bi,3} \tag{5.203}$$

$$\left.\frac{\partial \psi_{s,3}}{\partial x}\right|_{x=x_3} = 0 \tag{5.204}$$

At the source–channel ($i = 2$) and channel–drain boundaries ($i = 3$), we impose conditions of continuity of the surface potential and electric displacement:

$$\psi_{s,i-1}(x_{i-1}) = \psi_{s,i}(x_{i-1}) \tag{5.205}$$

$$\left.\frac{\partial \psi_{s,i-1}}{\partial x}\right|_{x=x_{i-1}} = \left.\frac{\partial \psi_{s,i}}{\partial x}\right|_{x=x_{i-1}} \tag{5.206}$$

Solution of these boundary conditions is not analytically possible, since Equations (5.201) to (5.206) lead to transcendental equations in the depletion region lengths. Therefore, we need to approximate the depletion region lengths.

This is done by applying the diode approximation to the source–channel and channel–drain depletion regions and by using the equations for the depletion region length in a diode:

$$L_1 = \sqrt{\frac{2\epsilon_{Si}|\psi_G - V_S - V_{bi,1}||N_2|}{q|N_1|(|N_2| + |N_1|)}} \tag{5.207}$$

$$L_3 = \sqrt{\frac{2\epsilon_{Si}|\psi_G - V_{DS} - V_{bi,2}||N_2|}{q|N_3|(|N_2| + |N_3|)}} \tag{5.208}$$

Using the above equations and those resulting from the boundary conditions (Equations (5.201) to (5.206), we get a system of linear equations in the coefficients C_i and D_i that we can solve to obtain the surface potential in each region. This procedure gives us the surface potential in the device while including the effects of the source and drain side depletion regions.

Note that while considering the source and drain side depletion regions, we did not include the effect of inversion charges. In the next section, we will study a method to include this effect by using smoothing functions.

5.7 Use of smoothing functions in the surface potential models

Smoothing functions are widely used in circuit simulations and are a method of mathematically switching from one regime of device operation to another.

Studies analysing the effect of inversion charges in TFETs have shown that the inversion charge concentration increases significantly after the gate potential ψ_G either (i) increases above the drain potential $\Psi_{drain} = V_{DS} + V_{bi,2}$ or (ii) decreases below the source potential $\Psi_{source} = V_S + V_{bi,1}$. Figure 5.23 shows the variation in the mid-channel potential of a TFET with varying gate bias. It can be observed that when $\Psi_{source} \leq \psi_G \leq \Psi_{drain}$, the inversion charge concentration is negligible, the mid-channel potential increases linearly with gate voltage and the slope is unity. However, when $\psi_G \geq \Psi_{drain}$ or $\psi_G \leq \Psi_{source}$, the inversion layer charge is significant. In this case, the increase in the channel potential with respect to the gate voltage is still linear, but with a slope reduced from 1 to $(1-\eta)$, where η is an empirical parameter [30].

To capture the surface potential variation as described above (Figure 5.23), a semi-empirical parameter called the "effective gate potential" $\psi_{G,eff}$ is used in place of the gate potential ψ_G to calculate the surface potential in the channel region. The behaviour of $\psi_{g,eff}$ is such that when $\Psi_{source} \leq \psi_G \leq \Psi_{drain}$,

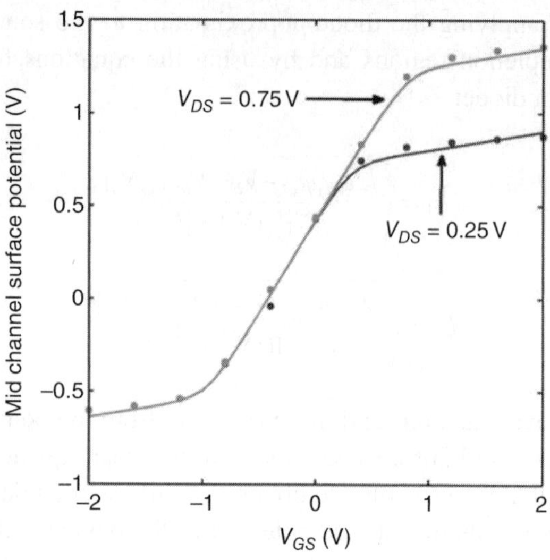

Figure 5.23 Variation in the mid-channel potential of an n-channel TFET with varying gate voltage (V_{GS}) [30]. Source: Reproduced with permission of IEEE.

$$\psi_{g,eff} = \psi_g \tag{5.209}$$

that is $\psi_{g,eff}$ increases linearly with ψ_g with a slope of 1. When $\psi_g \geq \Psi_{drain}$, pinning of the surface potential occurs on the drain side and the increase in $\psi_{g,eff}$ is now linear with a reduced slope of $(1-\eta)$:

$$\psi_{g,eff} = \eta \Psi_{drain} + (1-\eta)\psi_g \tag{5.210}$$

Similarly, when $\psi_g \leq \Psi_{source}$, pinning of the channel potential occurs on the source side and the increase in $\psi_{g,eff}$ is linear with a reduced slope $(1-\eta)$:

$$\psi_{g,eff} = \eta \Psi_{source} + (1-\eta)\psi_g \tag{5.211}$$

To model the transitions from Equations (5.210) to (5.211), a semi-empirical approach is adopted by using the following smoothing function:

$$\psi_{g,eff} = \psi_g - \eta\phi_t \ln\left(1 + e^{(\psi_g - \Psi_{drain})/\phi_t}\right) + \eta\phi_t \ln\left(1 + e^{(\psi_{source} - \psi_g)/\phi_t}\right) \tag{5.212}$$

where ϕ_t is a dimensionless empirical smoothing parameter obtained by fitting the simulated and the modelled transfer characteristics [15]. Physically, when the gate potential increases to a point where $(\psi_g - \Psi_{drain})$ is of the order of ϕ_t or more, there is an appreciable change in the slope of the surface potential, as shown in Figure 5.23. The exponential nature of the above function ensures the continuity and infinite differentiability of the surface potential, leading to the continuity and infinite differentiability of all the obtained characteristics. While the pseudo-2D method gives us results with reasonable accuracy and also provides analytical expressions for the surface potential, it is not able to predict the effects of inversion charges in a rigorous manner. In the next section, we will study the variational approach towards modelling the surface potential of a TFET, which predicts the effect of inversion charges in a TFET from *ab initio* theoretical principles rather than semi-empirical formulations.

References

[1] K. K. Young, "Short-Channel Effect in Fully Depleted SOI-MOSFETs", *IEEE Trans. Electron Devices*, vol. 40, no. 10, pp. 1812–1817, October 1993.

[2] M. J. Kumar and G. V. Reddy, "Diminised Short Channel Effects in Nanoscale Double-Gate Silicon-on-Insulator Metal Oxide Field Effect Transistors Due to Induced Back-Gate Step Potential", *Japanese Journal of Applied Physics*, vol. 44, no. 9A, pp. 6508–6509, September 2005.

[3] A. Chaudhry and M. J. Kumar, "Controlling Short-Channel Effects in Deep Submicron SOI MOSFETs for Improved Reliability: A Review", *IEEE Trans. on Device and Materials Reliability*, vol. 4, pp. 99–109, March 2004.

[4] J. Wan, C. L. Royer, A. Zaslavsky and S. Cristoloveanu, "A Tunneling Field Effect Transistor Model Combining Interband Tunneling with Channel Transport", *J. Appl. Phys.*, vol.110, no.10, pp. 104503–104503–7, 2011.

[5] A. Pan and C. O. Chui, "A Quasi-Analytical Model for Double-Gate Tunneling Field-Effect Transistors", *IEEE Electron Device Lett.*, vol. 33, pp. 1468–1470, October 2012.

[6] L. Zhang, X. Lin, J. He and M. Chan, "An Analytical Charge Model for Double-Gate Tunnel FETs", *IEEE Trans. Electron Devices*, vol. 59, no. 12, pp. 3217–3223, December 2012.

[7] A. Pan, S. Chen and C. O. Chui, "Electrostatic Modelling and Insights Regarding Multigate Lateral Tunneling Transistors", *IEEE Trans. Electron Devices*, vol. 60, no. 9, pp. 2712–3223, December 2012.

[8] L. Zhang and M. Chan, "SPICE Modelling of Double-Gate Tunnel-FETs Including Channel Transports", *IEEE Trans. Electron Devices*, vol. 61, no. 2, pp. 300–307, February 2014.

[9] R. Vishnoi and M. J. Kumar, "Compact Analytical Model of Dual Material Gate Tunneling Field Effect Transistor using Interband Tunneling and Channel Transport," *IEEE Transactions on Electron Devices*, vol. 61, no. 6, pp. 1936–1942, June 2014.

[10] R. Vishnoi and M. J. Kumar, "A Pseudo 2D-Analytical Model of Dual Material Gate All-Around Nanowire Tunneling FET", *IEEE Transactions on Electron Devices*, vol. 61, no. 7, pp. 2264–2270, July 2014.

[11] R. Vishnoi and M. J. Kumar, "Compact Analytical Drain Current Model of Gate-All-Around Nanowire Tunneling FET", *IEEE Transactions on Electron Devices*, vol. 61, no. 7, pp. 2599–2603, July 2014.

[12] R. Vishnoi and M. J. Kumar, "Two Dimensional Analytical Model for the Threshold Voltage of a Tunneling FET with Localized Charges", *IEEE Transactions on Electron Devices*, Vol. 61, no. 9, pp. 3054 – 3059, September 2014.

[13] R. Vishnoi and M. J. Kumar, "An Accurate Compact Analytical Model for the Drain Current of a TFET from Sub-threshold to Strong Inversion", *IEEE Transactions on Electron Devices*, vol. 62, no. 2, pp. 478–484, February 2015.

[14] R. Vishnoi and M. J. Kumar, "A Compact Analytical Model for the Drain Current of Gate All Around Nanowire Tunnel FET Accurate from Sub-threshold to ON-state", *IEEE Transactions on Nanotechnology*, vol. 14, no. 2, pp. 358–362, March 2015.

[15] P. Pandey, R. Vishnoi and M. J. Kumar, "A Full-Range Dual Material Gate Tunnel Field Effect Transistor Drain Current Model Considering Both Source and Drain Depletion Region Band-to-Band Tunneling", *Journal of Computational Electronics*, vol. 14, no. 1, pp. 280–287, March 2015.

[16] R. Vishnoi and M. J. Kumar, "Numerical Study of the Threshold Voltage of TFETs with Localized Charges", *IEEE International Conference on Emerging Electronics*, December 2014.

[17] P. Pandey, R. Vishnoi and M. J. Kumar, "Drain Current Model for SOI TFET Considering Source and Drain Side Tunneling", *IEEE International Conference on Emerging Electronics*, December 2014.

[18] R. Vishnoi and M. J. Kumar, "A Compact Analytical Model for the Drain Current of a TFET with Nonabrupt Doping Profile Incorporating the Effect of Band-gap Narrowing", *IEEE International Conference on Nanotechnology*, Rome, Italy, July 2015.

[19] R. Vishnoi and M. J. Kumar, "Two Dimensional Analytical Model for the Threshold Voltage of a Gate All Around Nanowire Tunneling FET with Localized Charges", *IEEE International Conference on Nanotechnology*, Rome, Italy, July 2015.

[20] C. Chen, S.-L. Ong, C.-H. Heng, G. Samudra and Y.-C. Yeo, "A Variational Approach to the Two-Dimensional Nonlinear Poisson's Equation for the Modelling of Tunneling Transistors", *IEEE Electron Device Letters*, vol. 29, pp. 1252–1255, November 2008.

[21] L. Komzisk, *Applied Calculus of Variations for Engineers*, CRC Press, 2009.

[22] X. Liang and Y. Taur, "A 2-D Analytical Solution for SCEs in DG MOSFETs", *IEEE Trans. on Electron Devices*, vol. 51, pp. 1385–1391, August 2004.

[23] E. Kreyszig, *Advanced Engineering Mathematics*, 8th edn, John Wiley & Sons, 1999.

[24] N. Cui, R. Liang, J. Wang and J. Xu, "Si-based Hetero-Material-Gate Tunnel Field Effect Transistor: Analytical Model and Simulation," *12th IEEE International Conference on Nanotechnology (IEEE-NANO)*, 2012, pp. 1–5.

[25] N. Cui, R. Lianga, J. Wang and J. Xu, "Two-dimensional analytical model of hetero strained Ge/strained Si TFET", *International Silicon–Germanium Technology and Device Meeting (ISTDM)*, 2012, pp. 1–2.

[26] M. Lee and W. Choi, "Analytical Model of Single-Gate Silicon-on-Insulator Tunneling Field-Effect Transistors", *Solid State Electron.*, vol. 63, no. 1, pp. 110–114, 2011.

[27] M. Gholizadeh and S.E. Hosseini, "A 2-D Analytical Model for Double-Gate Tunnel FETs", *IEEE Trans. on Electron Devices*, vol. 61, pp. 1494–1500, May 2014.

[28] L. Liu, D. Mohanta, and S. Datta, "Scaling Length Theory of Double-Gate Interband Tunnel Field-Effect Transistors", *IEEE Trans. Electron Devices*, vol. 59, no. 4, pp. 902–908, April 2012.

[29] S. Lin and J. Kuo, "Modeling the Fringing Electric Field Effect on the Threshold Voltage of FD SOI nMOS Devices with the LDD/Sidewall Oxide Spacer Structure", *IEEE Trans. Electron Devices*, vol. 50, no. 12, pp. 2559–2564, December 2003.

[30] K. Boucart and A. M. Ionescu, "Double-Gate Tunnel FET with High-*k* Gate Dielectric", *IEEE Trans. Electron Devices*, vol. 54, no. 7, pp. 1725–1733, July 2007.

6

Modelling the drain current

The problem of modelling the drain current in a TFET can be divided into two major parts – finding the surface potential and finding the tunnelling rate. The former has been discussed in the preceding chapter. In this chapter, we will outline the common methods used to find the tunnelling rate in a TFET and thus obtain the drain current in the device.

The band diagram of a TFET in the ON-state is shown in Figure 6.1(a) and the band diagram of the source–channel junction is shown in greater detail in Figure 6.1(b). In the ON-state of the TFET, most of the current is due to tunnelling across the source–channel junction, depicted by the horizontal arrow in Figure 6.1(b). We can consider this tunnelling from two perspectives – electrons at different energies E_i tunnelling from the point P_i in the source valence band to the point Q_i in the channel valence band, and particles at different positions x_i tunnelling from P_i to Q_i. The former approach usually corresponds to a non-local perspective of tunnelling, in which case the tunnelling probability is dependent on the potential in the entire region from the point P_i to the point Q_i. The latter approach usually applies a local perspective to tunnelling, with the tunnelling probability dependent only on the electric field at the initial position x_i.

It should also be noted that the tunnelling probability is different for every position x_i (or energy E_i, depending on our perspective). Therefore, to find the tunnelling rate throughout the device, we need to integrate the individual tunnelling probabilities obtained for every initial position (or energy). Since the functions describing the tunnelling probability are quite complicated, this integration

Tunnel Field-Effect Transistors (TFET): Modelling and Simulation, First Edition. Jagadesh Kumar Mamidala, Rajat Vishnoi and Pratyush Pandey.

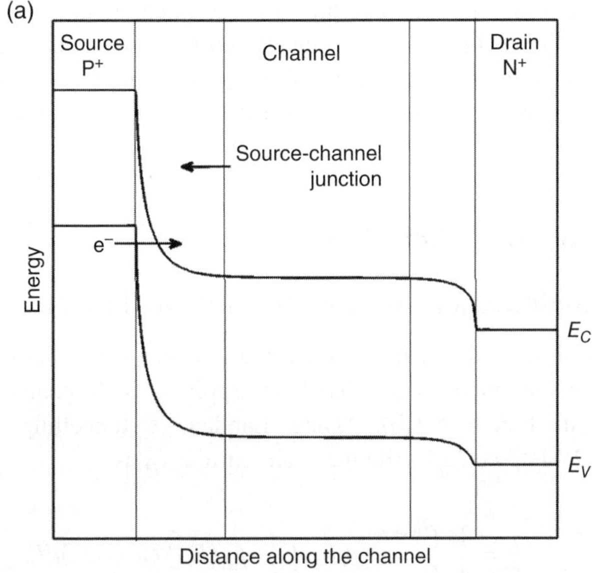

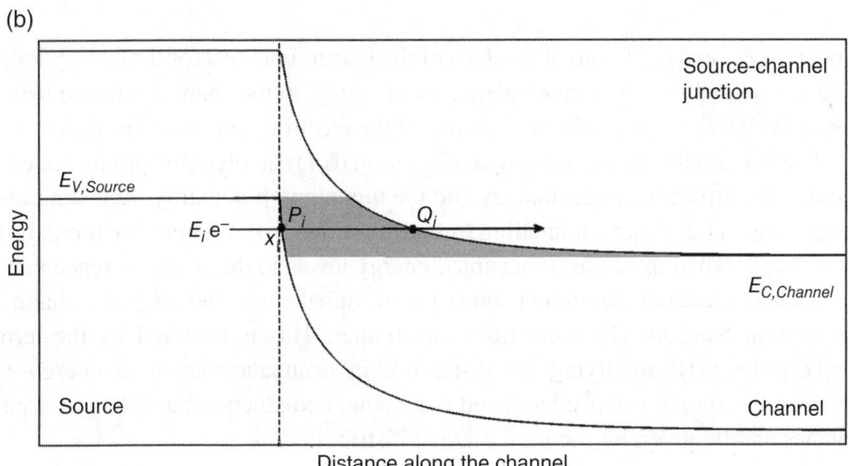

Figure 6.1 (a) Band diagram of an n-channel TFET at the surface in the ON-state. (b) Band diagram in the source–channel junction.

may not have a closed form solution, and evaluating it often requires various approximations.

In this chapter, we will discuss the local and non-local methods used to find the tunnelling probability in TFETs. Corresponding to each method, we will first obtain an expression for the tunnelling probability from the surface potential.

We will next study various techniques used to integrate the tunnelling probabilities. We would then be able to calculate the current for a given biasing and obtain the transfer and output characteristics of the device. Finally, using the device characteristics thus obtained, we will discuss different definitions of the threshold voltage of a TFET.

6.1 Non-local methods

6.1.1 Landauer's tunnelling formula in TFETs

The non-local method considers tunnelling to occur at an energy level E_i, from the point P_i in the source valence band to the point Q_i in the channel valence band, as shown in Figure 6.1(b). Using Landauer's tunnelling formula [1–6] (Equation (2.44)), we can write the drain current I_D as

$$I_D = \frac{2q}{\hbar} \int_{E_{C,Channel}}^{E_{V,Source}} (f_{P_i}(E_i) - f_{Q_i}(E_i)) T_{BTBT}(E_i) dE_i \tag{6.1}$$

where $f_{P_i}(E_i)$ and $f_{Q_i}(E_i)$ are the values of the Fermi functions for the energy level E_i at the point P_i in the source valence band and Q_i in the channel valence band, respectively; $T_{BTBT}(E_i)$ is the probability of the electron tunnelling from P_i to Q_i.

The integral for the device current (Equation (6.1)) involves the product of two terms – the difference in occupancy and the tunnelling probability. Recalling our discussion on Landauer's tunnelling formula (Section 2.3), we note that tunnelling occurs only when there are (i) occupied energy levels in the source valence band *from* where electrons can tunnel and (ii) unoccupied energy levels in the channel conduction band *to* which electrons can tunnel. This is captured by the term $(f_{P_i}(E_i) - f_{Q_i}(E_i))$, signifying the difference in occupancy levels at energy E_i between the source valence band and the channel conduction band. These occupancies can be given by the Fermi–Dirac distribution as

$$f_{P_i}(E_i) = \frac{1}{1 + \exp((E_i - E_{F,Source})/kT)} \tag{6.2}$$

$$f_{Q_i}(E_i) = \frac{1}{1 + \exp((E_i - E_{F,Channel})/kT)} \tag{6.3}$$

where $E_{F,Source}$ and $E_{F,Channel}$ are the Fermi energy levels in the source and drain, respectively. The integration (6.1) is carried out for all energy levels from which tunnelling can take place, starting from the valence band edge of the source $E_{V,Source}$ to the conduction band edge of the channel $E_{C,Channel}$.

To evaluate the integral (6.1), we need to find the tunnelling probability $T_{BTBT}(E_i)$. In non-local models, the most widely used method to calculate the tunnelling probability is the WKB approximation, which we will study in the following section.

6.1.2 WKB approximation in TFETs

In Section 2.2, we learnt the WKB approximation, which enabled us to calculate the tunnelling probability for a general potential barrier. We will now use the WKB approximation to find the probability of an electron in the valence band of the source tunnelling to the conduction band of the channel. From Equations (2.28) and (2.30), we can write the tunnelling rate T_{BTBT} given by the WKB approximation as

$$T_{BTBT} = e^{-2\gamma} \tag{6.4}$$

$$\gamma = \int_{x_{P_i}}^{x_{Q_i}} \left| \sqrt{\frac{2m(V(x)-E)}{\hbar^2}} \right| dx \tag{6.5}$$

In the preceding chapter, we studied different methods used to obtain the surface potential in the device if the biasing is known. We will now study how the surface potential $\psi_s(x)$ can be converted into the potential barrier $V(x)$ of the tunnelling problem.

Let us examine the band diagram of the device in the ON-state, focusing our attention to the region in which tunnelling is taking place, as shown in Figure 6.2. Since $E/q = \psi$, the shape of the surface potential gives us the shape of the energy band edges of the material. Hence, if we consider $q\psi_s(x)$ to be the valence band energy, the plot of the conduction band would be given by $E_g + q\psi_s(x)$, where E_g is the bandgap of the material. Examining the tunnelling path taken by an electron at an arbitrary position P_i and energy E_i in the valence band of the source, the potential barrier that the electron encounters (the shaded shape in Figure 6.2) is given as

$$V(x) = E_g + q\psi_s(x) \tag{6.6}$$

and this barrier $V(x)$ extends from P_i to Q_i. Therefore, the term $(V(x) - E)$ in the WKB approximation (Equation (6.5)) is

$$V(x) - E = V(x) - E_i = E_g + q\psi_s(x) - E_i \tag{6.7}$$

Our problem is not completely solved yet – while the integrand of Equation (6.5) is known, we need to obtain the limits of integration x_{P_i} and

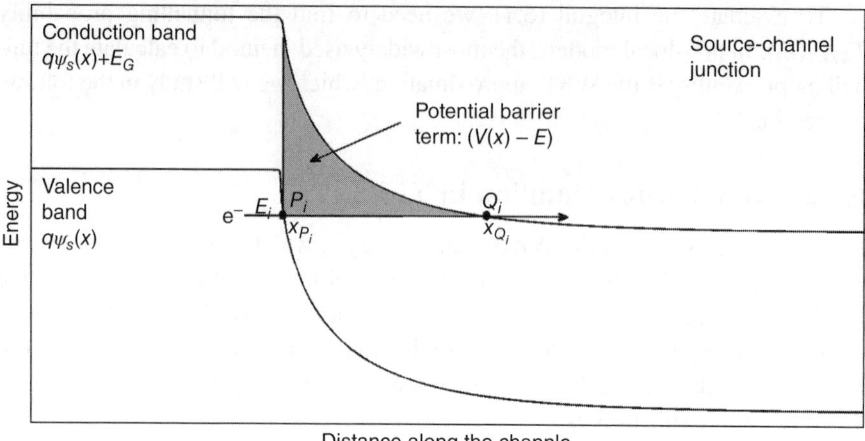

Figure 6.2 Band diagram in the source–channel junction of the TFET in the ON-state showing the tunnelling of an electron across the bandgap.

x_{Q_i}. Since we have considered $q\psi_s(x)$ to represent the valence band energy, at the position x_{P_i}, we must have $q\psi_s(x_{P_i}) = E_i$. At the final point Q_i, the conduction band energy equals the initial energy E_i. Since the conduction band energy is given as $E_g + q\psi_s(x)$, we can find the position x_{Q_i} by solving

$$q\psi_S(x_{Q_i}) + E_g = E_i = q\psi_s(x_{P_i}) \tag{6.8}$$

We now have the shape and width of the potential barrier for an electron at any arbitrary energy E_i and we can find the tunnelling probability $T(E_i)$ for this energy if the surface potential in the device $\psi_s(x)$ is known. To better understand this process, let us now find the current in the device after assuming a particular form of the surface potential.

6.1.3 Obtaining the drain current

Till this point, we have written expressions to find the tunnelling probability and the current for a general surface potential $\psi_s(x)$. However, due to the complicated integrals involved (Equations (6.1) and (6.5)), evaluation of the current is very difficult without using approximations.

A widely used method to solve this problem involves two approximations – the potential in the tunnelling region is linear and the tunnelling probability $T(E_i)$ in Equation (6.1) is constant. The first approximation simplifies the process of obtaining the tunnelling probability in Equation (6.4). The second approximation

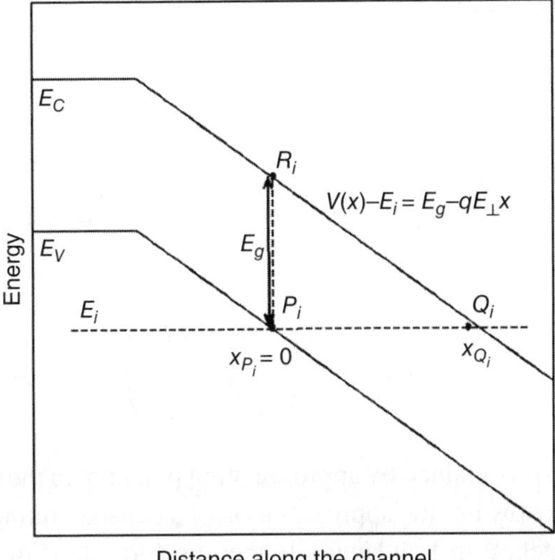

Figure 6.3 Tunnelling of an electron across in a linear potential profile.

brings the term $T(E_i)$ out of the integral in Equation (6.1), making it easier to evaluate the integral and obtain the current.

Let us first find the tunnelling probability $T(E_i)$ when the potential in the tunnelling region is linear, as shown in Figure 6.3. In this case, the tunnelling barrier is the triangle $P_iQ_iR_i$. The side R_iQ_i of this triangle represents the term $(V(x)-E)$ of Equation (6.5). Since the potential is linear, the longitudinal electric field $E_\perp$ in the tunnelling region is constant. Taking the origin at the initial position P_i (i.e. $x_{P_i}=0$), and using the fact that the surface potential is linear with a slope $-qE_\perp$, we can rewrite Equation (6.7) as

$$V(x)-E=E_g-qE_\perp x \tag{6.9}$$

Using Equation (6.9) in Equation (6.8), we get the position x_{Q_i} of the final point Q_i as

$$x_{Q_i}=E_g/qE_\perp \tag{6.10}$$

We can now rewrite the WKB integral (Equation (6.5)) using the above Equations (6.9) and (6.10):

$$\gamma=\int_0^{E_g/qE_\perp}\left|\sqrt{\frac{2m(E_g-qE_\perp x)}{\hbar^2}}\right|dx \tag{6.11}$$

Evaluating Equation (6.11) gives

$$\gamma = \frac{2\sqrt{2mE_g^3}}{3q\hbar E_\perp} \tag{6.12}$$

Substituting the value of γ in Equation (6.4), the tunnelling probability can be written as

$$T_{BTBT} = \exp\left(-\frac{4\sqrt{2mE_g^3}}{3q\hbar E_\perp}\right) \tag{6.13}$$

Equation (6.13) is obtained by approximating potential in the tunnelling region to be linear. Let us now use the approximation of a constant tunnelling probability to evaluate the integral in Landauer's tunnelling formula (Equation (6.1)). The variable of integration is the energy E_i, and the only functions of energy within the integral correspond to the occupancy levels in the source ($f_{P_i}(E_i)$) and in the channel ($f_{Q_i}(E_i)$) given by Equations (6.2) and (6.3). Substituting Equations (6.2), (6.3) and (6.13) in Equation (6.1) and evaluating the resulting integral, we get the drain current I_D as

$$I_D = \frac{2qkT}{\hbar} T_{BTBT} \ln\left|\frac{(1+\exp(E_{F,Source}-E_{C,Channel}/kT))(1+\exp(E_{F,Channel}-E_{V,Source}/kT))}{(1+\exp(E_{F,Source}-E_{V,Source}/kT))(1+\exp(E_{F,Channel}-E_{C,Channel}/kT))}\right| \tag{6.14}$$

It should be noted that while developing this drain current model, the longitudinal electric field $E_\perp$ is considered to be constant. However, the surface potential model would usually result in a varying electric field in the tunnelling region. Therefore, many different approximations are used to find the appropriate value of $E_\perp$ in Equation (6.13). Some methods take the maximum electric field in the tunnelling region to be $E_\perp$, but this leads to overestimation of the drain current. This problem can be solved by taking the average of the minimum and maximum electric field or by calculating the average electric field in the tunnelling region and taking it to be $E_\perp$. This approach also has a drawback, since the contribution of a higher electric field to the tunnelling current is far greater than that of a lower electric field due to the exponential dependence of the band-to-band generation rate on the electric field (Equation (6.13)). Another commonly used approach is to write $E_\perp$ in terms of the screening length λ of the device:

$$E_\perp = \frac{E_G + (E_{V,Source} - E_{C,Channel})}{q\lambda} \tag{6.15}$$

As described in Section 2.4, a completely non-local approach is more accurate than a local approach to find the drain current of a TFET. However, the approximations that we have made in order to obtain an analytical result for the drain current lead to certain disadvantages. The non-local method is able to effectively capture the effect of the density of states and occupancy levels in the source and the channel; it will be clear from the next section that local methods are not able to accomplish this. On the other hand, due to the inclusion of terms relating to the density of states, the non-local methods predict a zero current before there is any overlap between the source and channel bands. Experimental results show that this is not the case and TFETs do have a non-zero subthreshold current. Finally, due to the approximation of a constant electric field (linear potential), this method is not able to incorporate the effects of the varying electric field within the channel. The local methods to calculate the current in a TFET that we will study in the next section provide a suitable alternative – they are able to predict the subthreshold current and can incorporate the effects of the varying electric field throughout the device.

6.2 Local methods

In the previous section, we studied the non-local approach to tunnelling, which considers electrons at different energies E_i tunnelling from the source valence band to the channel conduction band. Due to this perspective, the variable of integration in the case of a non-local approach (Equation (6.1)) was energy. We will now study the local approach for modelling the drain current in a TFET, which considers electrons at different points x_i tunnelling from the source valence band to the channel conduction band. In this case, to obtain the drain current I_D, the variable of integration is volume:

$$I_D = q\int G_{btb} dV \tag{6.16}$$

where G_{btb} is the band-to-band generation rate at the position of the volume element dV and is usually obtained by using Kane's model [7] (Section 2.4.2.1):

$$G_{btb} = \frac{E^{2.5} m_r^{1/2}}{18\pi\hbar^2 E_G^{1/2}} \exp\left\{\frac{-\pi m_r^{1/2} E_G^{3/2}}{2\hbar|E|}\right\} = A\frac{E^{2.5}}{E_G^{1/2}} \exp\left\{-B\frac{E_G^{3/2}}{|E|}\right\} \tag{6.17}$$

The above expression gives the band-to-band generation rate at any point in the device in terms of the local electric field E at that point. Integrating this generation rate over the entire device volume gives the total number of carriers generated due to tunnelling. As explained in (Section 2.4.2.1), it is assumed that the drain bias sweeps all the generated carriers towards the drain. Therefore, the drain current is equal to the rate of the total number of carriers generated due to tunnelling, and is given by integrating G_{btb} throughout the device volume.

However, integrating G_{btb} throughout the device may not be the most computationally efficient method to obtain the drain current. Observing the band diagram along the surface of a TFET in the ON-state (Figure 6.1(a)), we can see that electrons are tunnelling only in the depletion region of the source–channel junction (which is termed the tunnelling region). Hence, to obtain the drain current, it is sufficient to integrate G_{btb} only in the tunnelling region.

The expression for G_{btb} (Equation (6.17)) shows that it is a function of the electric field at a given point. Therefore, if the potential in the device is known, its derivative will give us the field, which can be subsequently used in Equation (6.17) to calculate G_{btb} at every point. We can then integrate G_{btb} over the tunnelling region to obtain the drain current. However, the expression for G_{btb} has polynomial as well as exponential dependency on the electric field. Due to this, in most cases there is no closed form integration of G_{btb}. Therefore, if an analytical expression for the drain current is required, we need to explore different approximations to evaluate this integral. We will now discuss different methods for finding the integral of G_{btb} over the volume of the tunnelling region.

6.2.1 Numerical integration

The most straightforward way of finding the integral given in Equation (6.16) is to evaluate it numerically [8–11]. We divide the length of the tunnelling region into extremely small steps (Δx) of x and take the summation of $G_{btb}(x)\Delta x$:

$$I_D = q\sum_{n=0}^{0} G_{btb}(x_n)\Delta x; \quad n = 1, 2, 3, \ldots, N \tag{6.18}$$

where N is a large integer. This gives us the most accurate value of the total tunnelling generation rate (G_{btb}) in the tunnelling region. However, this summation involves a large number of steps and requires a huge amount of computational resources, making it highly inefficient for circuit simulation. Moreover, this method does not give us an analytical expression for the drain current of a TFET. We will now move to different methods that approximate the integration of G_{btb} and give us an analytical expression for the drain current.

6.2.2 Shortest tunnelling length

In the previous section, the tunnelling generation rate was integrated numerically, which is an accurate but computationally inefficient method of obtaining the drain current. More importantly, as it does not give an analytical expression, it provides little insight into the process of tunnelling in the device. Hence, it is very important to develop methods that are compact and analytical, so as to provide faster computation and better insights into the device.

To get a closed-form expression of the tunnelling generation rate (G_{btb}), we need to approximate the integration given in Equation (6.16). One way to approximate the integral is to exploit the exponential dependence of G_{btb} on the electric field. As G_{btb} is an exponential function of the electric field, it dies out very quickly with a decrease in the electric field. Hence, we can calculate G_{btb} at the region of maximum electric field and neglect the G_{btb} in the rest of the tunnelling region [12]. Let us now have a look at Figure 6.4, which shows the surface potential in a TFET in the ON-state. As can be seen, the surface potential curve is steepest at the source–channel junction and hence has the maximum electric field at the source–channel junction. However, since tunnelling is a process that takes place across a tunnelling width, we use the average electric field (E_{TW}) over the shortest tunnelling length (L_{TW}) to find G_{btb} from Equation (6.15) and multiply it with a constant so as to find the drain current.

The shortest tunnelling length (L_{TW}) is the distance from the source–channel junction over which the surface potential drops by E_g/q in a p-TFET and vice versa in an n-TFET, as shown in Figure 6.4. In other words, the shortest tunnelling length is the shortest distance over which we get an alignment between the conduction band in the source and the valence band in the channel. Hence, L_{TW} can be written as

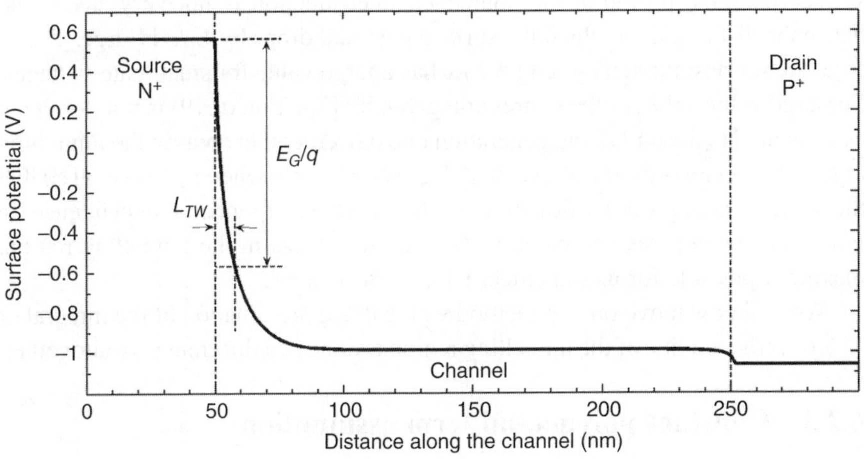

Figure 6.4 Surface potential of a p-channel TFET in the ON-state.

$$L_{TW} = x[\psi(source\ - E_G/q)] - x[\psi(source)] \tag{6.19}$$

The average electric field (E_{TW}) over the shortest tunnelling length (L_{TW}) is given by

$$E_{TW} = \frac{E_G}{qL_{TW}} \tag{6.20}$$

Now the drain current can be approximated to be proportional to G_{btb} given by Equation (6.15) with $E = E_{TW}$:

$$I_D \propto A \times \frac{E_{TW}^{2.5}}{E_G^{0.5}} \times \exp\left[-B\frac{E_G^{1.5}}{E_{TW}}\right] \tag{6.21}$$

Since the tunnelling of carriers is taking place in the entire thickness of the inversion layer, as shown in Figure 4.1, we have to multiply the above expression by a constant factor A_1, which incorporates the thickness of the inversion layer:

$$I_D = A_1 \times A \times \frac{E_{TW}^{2.5}}{E_G^{0.5}} \times \exp\left[-B\frac{E_G^{1.5}}{E_{TW}}\right] \tag{6.22}$$

The above model (6.20), is a closed-form expression for the drain current of the TFET. It can use any of the surface potential models discussed in Chapter 5 to find L_{TW} and then use it in Equations (6.17) to (6.20) to give us the drain current. The above model, however, is not very accurate, especially for low gate voltages when the shortest tunnelling length is not very small. At small gate voltages, the surface potential profile at the source–channel junction is not very steep, and hence the distance over which the surface potential drops by E_g/q is large. Therefore, the shortest tunnelling length L_{TW} has a large value for small gate voltages. For large values of L_{TW}, the expression given by Equation (6.19) is not dominant as compared to the tunnelling generation rate (G_{btb}) at other areas in the tunnelling region. In other words, the above model is accurate only when L_{TW} is small so that the value of G_{btb} given by Equation (6.16) due to its exponential dependence on E_{TW} becomes large as compared to G_{btb} at other areas in the tunnelling region, making it possible for us to neglect G_{btb} in those areas.

We will now move on to a method where the approximation of the integral of G_{btb} over the volume of the tunnelling region is done in a different way altogether.

6.2.3 Constant polynomial term assumption

In the previous section, we discussed a method of finding the integration of G_{btb}, which used the exponential dependence of G_{btb} on the electric field for approximating the integration. In this section, however, we will discuss a different approach for approximating the integral given by Equation (6.16) [13–15].

Let us rewrite Equation (6.17) as

$$G_{btb} = \frac{E^{2.5} m_r^{1/2}}{18\pi\hbar^2 E_G^{1/2}} \exp\left\{ \frac{-\pi m_r^{1/2} E_G^{3/2}}{2\hbar|E|} \right\} = A \frac{E^{2.5}}{E_G^{1/2}} \exp\left\{ -B \frac{E_G^{3/2}}{|E|} \right\} \tag{6.23}$$

Here we can see that G_{btb} has both exponential and polynomial terms in the electric field, which makes it impossible to get a closed-form expression for the integration of G_{btb}. However, as the polynomial term is expected to change more slowly with the electric field, as compared to the exponential term, we can assume the polynomial term $E(x)^{2.5}$ to be constant in the tunnelling region and integrate only in the exponential term. This gives us the following equation:

$$I_D = \int G_{btb} dx = \int A \frac{E(x)^{2.5}}{E_G^{1/2}} \exp\left\{ -B \frac{E_G^{3/2}}{|E(x)|} \right\} dx = A \frac{E(x_0)^{2.5}}{E_G^{1/2}} \int \exp\left\{ -B \frac{E_G^{3/2}}{|E(x)|} \right\} dx \tag{6.24}$$

where the integration is carried out over the length of the tunnelling region. The position $x = x_0$ at which the constant value of the polynomial term is chosen can either be at the source–channel interface (position of the maximum electric field) or at any suitable average value. The expression for $E(x)$ is given by using any of the surface potential models discussed in Chapter 5, which would enable us to write the integral in terms of x. The final form of expression of Equation (6.24) will, therefore, depend on the surface potential model used. As this model takes into account the varying electric field by incorporating it into the integral, it intuitively appears to be more accurate than the shortest tunnelling length approach, where only a constant electric field was used in the expression (Equation (6.21)) to calculate the device current. However, for steeper potential profiles at a high gate bias, the variation in the polynomial term $E(x_0)^{2.5}$ in Equation (6.24) may also be significant, weakening the approximation we have made here and reducing the accuracy. Despite this shortcoming, this approach of approximation of the integral will be more accurate than the shortest tunnelling length method.

The approaches for approximating the integral of the tunnelling generation rate that we have discussed so far have limited accuracy. Even after proper calibration of the constant terms A and B (6.23), these models fail to predict the drain current accurately in the entire range of gate voltages. Hence, it is necessary to develop a unified approach that gives us a compact (or closed-form) expression for the drain current of a TEFT that is accurate in the entire range of gate voltages.

We will now move on to the next approach of approximating the integral of the tunnelling generation rate. This approach is graphical and makes use of the differential (or the slope) of the expression given in Equation (6.17) to evaluate the integral of G_{btb}, which would give us the total tunnelling generation rate in the device.

6.2.4 Tangent line approximation

The methods for approximating the integral of G_{btb} discussed so far have limited accuracy. We will now discuss a method that, along with being compact and analytical, can also achieve a high level of accuracy in evaluating the integral of G_{btb} in the entire range of gate voltages. We will integrate the tunnelling generation rate G_{btb} for the TFET shown in Figure 4.1, which is in the ON-state ($V_{GS} = -3$ V and $V_{DS} = -0.5$ V). The variation of G_{btb} along the length of the channel in the tunnelling region is shown in Figure 6.5. As can be seen in the figure, G_{btb} is highest at the source–channel junction and decreases steeply to a negligible value within a few nanometres along the channel. The decrease is steepest at the source–channel junction and becomes less steep as we move along the channel. This nature of G_{btb} can be utilised to draw tangents at appropriate points on its graph. We then find the area under these tangents, giving us a good approximation of the integration of G_{btb}. This method is therefore called the tangent line approximation method

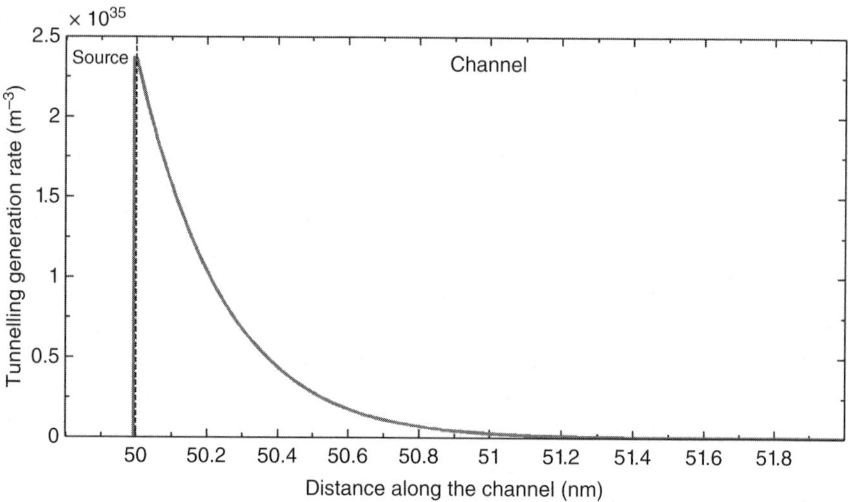

Figure 6.5 Tunnelling generation rate (G_{btb}) at the surface along the channel of a TFET starting from the source–channel junction.

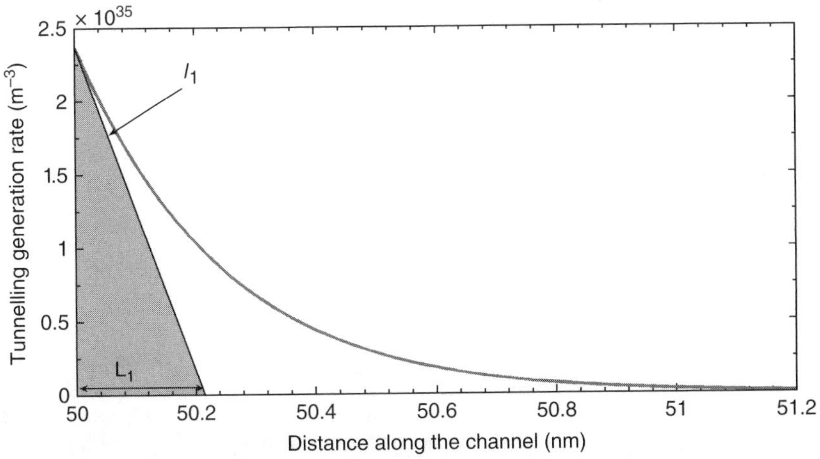

Figure 6.6 Tangent line l_1 to the G_{btb} curve. The shaded area is G_1.

[16, 17]. Let us now see how this method works. For simplicity, we will define the origin (i.e. $x=0$) at the source–channel junction.

We first start with drawing a tangent to the G_{btb} curve at $x=0$, that is the source–channel junction where G_{btb} has the maximum value, and call it line l_1, as shown in Figure 6.6. We extend this tangent line (l_1) to the point where it intersects the x-axis ($x=L_1$) and find the area G_1 under it (shaded region shown in Figure 6.6). Since the line l_1 is a tangent to the G_{btb} curve at $x=0$, its slope is equal to $G'_{btb}(0)$ and hence we get the following expressions for L_1 and G_1:

$$L_1 = G_{bib}(0)/G'_{bib}(0) \tag{6.25}$$

$$G_1 = 0.5G'_{bib}(0)L_1^2 \tag{6.26}$$

At the x-intercept of the line l_1 (i.e. $x=L_1$), we draw another tangent line (line l_2) to the G_{btb} curve and extend it to the point where it intersects the x-axis ($x=L_1+L_2$) and to the point where it meets the previous tangent line $l_2(x=L_1+L_2-L_{1d})$, as shown in Figure 6.7. We now find the area G_2 under this line (l_2). We also find the area G_{1d} common under the lines l_1 and l_2 (shaded region shown in Figure 6.8). The slope of line l_2 is $G'_{btb}(L_1)$ and we get the following expressions:

$$L_2 = G_{bib}(L_1)/G'_{bib}(L_1) \tag{6.27}$$

$$L_{1d} = G'_{bib}(0)L_2/\left[G'_{bib}(0)-G'_{bib}(L_1)\right] \tag{6.28}$$

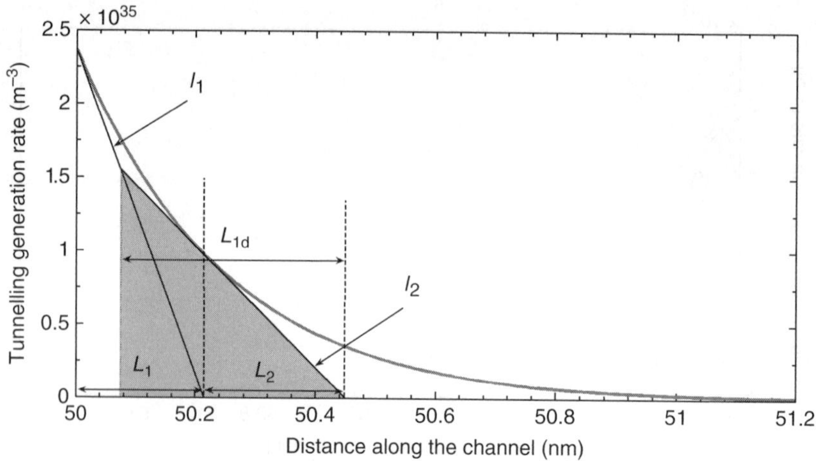

Figure 6.7 Tangent line l_2 to the G_{btb} curve. The shaded area is G_2.

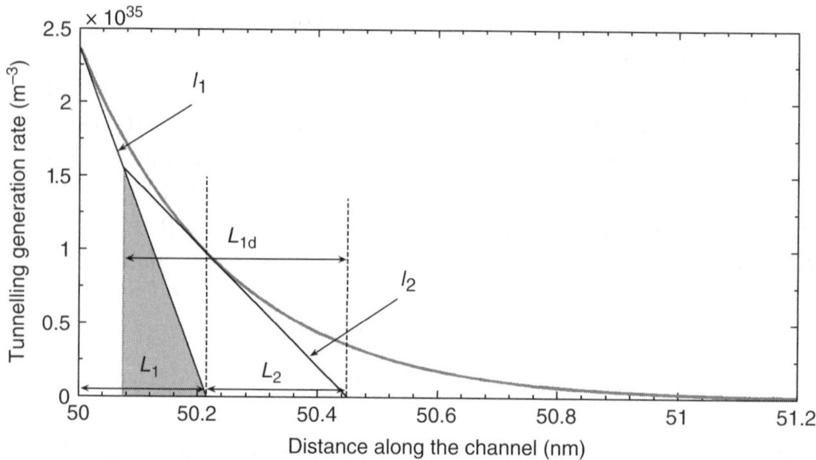

Figure 6.8 Tangent line l_3 to the G_{btb} curve. The shaded area is G_3.

$$G_2 = 0.5G'_{bib}(L_1)L_{1d}^2 \tag{6.29}$$

$$G_{1d} = 0.5G'_{bib}(0)(L_{1d} - L_2)^2 \tag{6.30}$$

We will now repeat the process outlined above once again. At the x-intercept of line l_2 (i.e. $x = L_1 + L_2$), we draw a tangent line (line l_3) to the G_{btb}

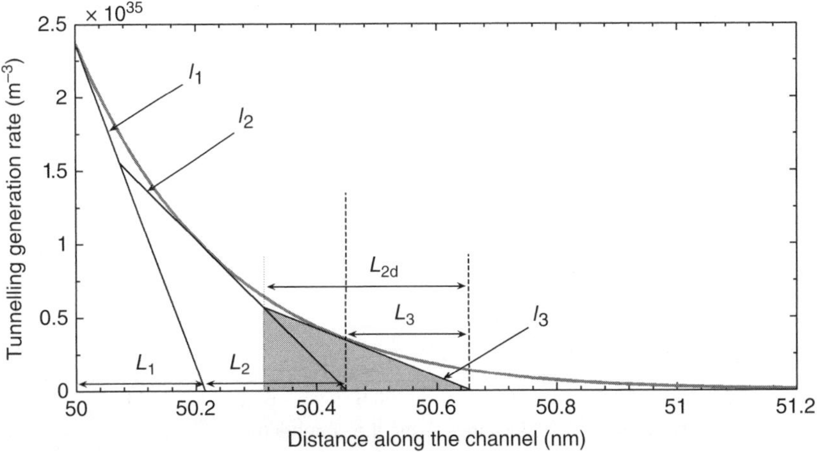

Figure 6.9 *Area common to the tangent lines l_1 and l_2, that is G_{1d}.*

curve and extend it to the point where it intersects the x-axis ($x = L_1 + L_2 + L_3$) and to the point where it meets line l_2 ($x = L_1 + L_2 + L_3 - L_{2d}$), as shown in Figure 6.9. We find the area G_3 under the line l_3 (shaded region shown in Figure 6.9) and also find the area G_{2d} common under the lines l_2 and l_3 (shaded region shown in Figure 6.10). The slope of line l_3 is $G'_{btb}(L_1 + L_2)$ and we get the following expressions:

$$L_3 = G_{bib}(L_1 + L_2)/G'_{bib}(L_1 + L_2) \tag{6.31}$$

$$L_{2d} = \frac{G'_{bib}(L_1)L_3}{\left[G'_{bib}(L_1) - G'_{bib}(L_1 + L_2)\right]} \tag{6.32}$$

$$G_3 = 0.5G'_{bib}(L_1 + L_2)L_{2d}^2 \tag{6.33}$$

$$G_{2d} = 0.5G'_{bib}(L_1)(L_{2d} - L_3)^2 \tag{6.34}$$

Finally, we add the areas under the three lines l_1, l_2 and l_3. Note that the areas common under any two lines have been added twice. Hence, we subtract the areas common between any two lines, giving us the total generation rate G_T:

$$G_T = G_1 + G_2 + G_3 - G_{1d} - G_{2d} \tag{6.35}$$

G_T gives us an approximate value of the integration of the tunnelling generation rate G_{btb}. The accuracy of G_T can be improved further by repeating the

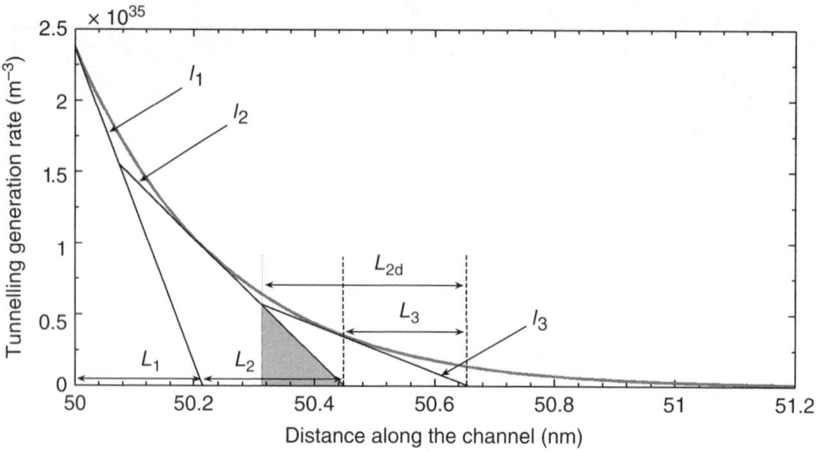

Figure 6.10 Area common to the tangent lines l_2 and l_3, that is G_{2d}.

process of drawing a tangent line and then finding the area under it, a greater number of times. For the nth repetition step, we will get the following set of equations:

$$L_n = \frac{G_{btb}(L_1 + L_2 + L_3 + \cdots + L_{n-1})}{G'_{btb}(L_1 + L_2 + L_3 + \cdots + L_{n-1})} \tag{6.36}$$

$$L_{n-1d} = \frac{G'_{btb}(L_1 + L_2 + \cdots + L_{n-2})L_n}{\left[G'_{btb}(L_1 + L_2 + \cdots + L_{n-2}) - G'_{btb}(L_1 + L_2 + \cdots + L_{n-1})\right]} \tag{6.37}$$

$$G_n = 0.5G'_{btb}(L_1 + L_2 + \cdots + L_{n-1})L^2_{n-1d} \tag{6.38}$$

$$G_{n-1d} = 0.5G'_{btb}(L_1 + L_2 + \cdots + L_{n-2})(L_{n-1d} - L_n)^2 \tag{6.39}$$

At the end of the nth step, the expression for G_T is given as follows:

$$G_T = G_1 + G_2 + \cdots + G_n - G_{1d} - G_{2d} - \cdots - G_{n-1d} \tag{6.40}$$

The drain current I_d can now be given as

$$I_d = qA_K G_T \tag{6.41}$$

where A_k is a constant, which incorporates the thickness t_{inv} of the inversion layer and is given by

$$A_K = \frac{A \times t_{inv}}{E_g^{0.5}} \tag{6.42}$$

where A is the pre-exponential constant of Kane's model (Equation (6.17)).

Table 6.1 Accuracy of tangent line approximation

Number of repetition steps	Accuracy
1	53%
2	80%
3	89%
4	92%
5	93%
6	93.6%
7	93.7%

We will now analyse the accuracy of the tangent line approximation method with a varying number of repetition steps. The accuracy of the tangent line approximation can be evaluated by dividing G_T (Equation (6.40)) by the numerically evaluated value of the integration of G_{btb} (x), given in Figure 6.5. In Table 6.1, we show the accuracy of the tangent line approximation with the number of repetition steps n. As we can see here, the accuracy settles to around 93% after the 4th step. Hence, using 5 to 6 repetition steps gives us very good accuracy of the integration of G_{btb}.

The method discussed in this section has closed-form equations and evaluates the integral of the tunnelling generation rate with more than 93% accuracy. Hence, by utilising the fact that any given function has a closed-form expression for its differential, we have developed a method for finding the drain current of a TFET that is computationally efficient and accurate at the same time. The final expressions for the drain current are closed-form equations and no iterations or numerical integrations are required at any step, making this method very useful for developing TFET models for circuit simulations.

6.3 Threshold voltage models

One of the most important parameters of a MOSFET is its threshold voltage. For a MOSFET, there is a clear physical definition of the threshold voltage – it is the gate voltage at which the magnitude of the conducting inversion layer charge in the channel is equal to the background concentration. However, since the current in a TFET is not controlled by the formation of a conducting inversion layer, but by modulation of the tunnelling barrier, this definition cannot be applied to TFETs.

Another point of dissimilarity from MOSFETs is that TFETs have two threshold voltages – the gate threshold voltage and the drain threshold voltage. The reason for this becomes clear if we remember our previous discussion

(Section 3.2.2.2) on the output characteristics of a TFET. We had studied that even if a high gate bias is applied, the TFET is in the OFF-state in the case of a low drain bias, as the channel potential is pinned to the drain potential. As the drain bias is increased, the TFET transitions from the OFF-state to the ON-state, thereby leading to a second threshold voltage, the drain threshold voltage.

Due to these reasons, the definitions of the threshold voltage used for a MOSFET could not be directly applied to TFETs. New definitions of the threshold voltage were needed for TFETs, some of which will be studied in this section.

6.3.1 Constant current method

This was one of the first methods used to extract the threshold voltage of a TFET and was based on practical considerations. The ON-state current of a TFET lies in the order of $1\ 10^{-6}\,A/\mu m$. Drawing parallels with MOSFETs, the current at the threshold voltage should be an order of magnitude lower than the ON-state current. Therefore, in this method, the threshold voltage is defined as that gate (or drain) voltage at which the current in the device $I_{DS} = 10^{-7}\,\mathrm{A}/\mu\mathrm{m}$.

When extracted using this method, the gate threshold voltage displays negligible dependence on the drain voltage (if $V_{DS} > 0.1\,\mathrm{V}$), as can be seen in Figure 6.11. Apart from this, as shown in Figure 6.12, there is a small amount of threshold voltage roll-off at low gate lengths.

The drain threshold voltage extracted by this method decreases with increasing gate voltage (Figure 6.13). This is because a higher gate voltage would lead to a greater electric field at the source–channel junction, even in case of pinning, leading to a higher current.

One variation of this method used in analytical modelling is the constant tunnelling length method, which we will now study.

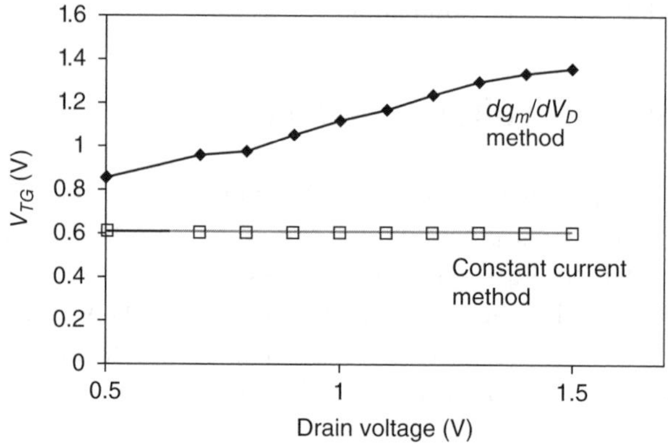

Figure 6.11 Dependence of the gate threshold voltage on the drain voltage for an n-channel TFET.

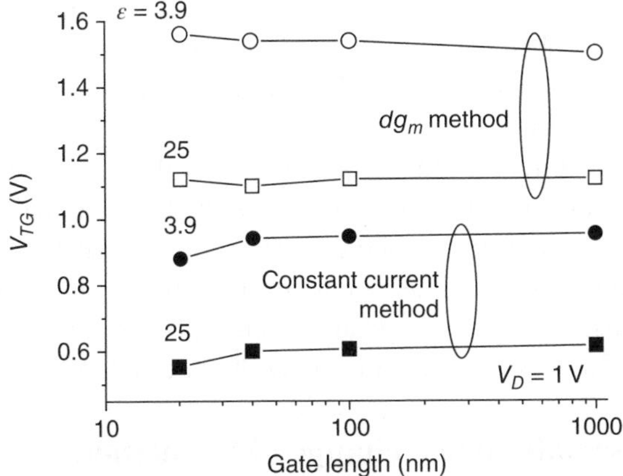

Figure 6.12 Variation in the gate threshold voltage with the gate length for an n-channel TFET.

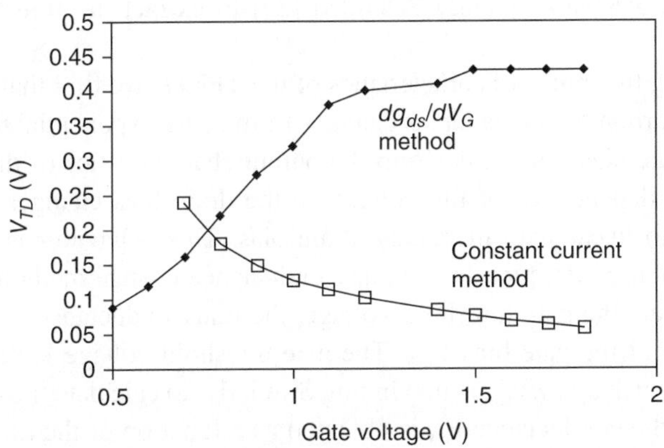

Figure 6.13 Dependence of the drain threshold voltage on the gate voltage for an n-channel TFET.

6.3.2 Constant tunnelling length

As described in Section 6.2.3, the drain current in a TFET is often written as a function of the tunnelling length (Equation (6.20)). For a particular device structure, the tunnelling length at which the current $I_{DS} = 10^{-7}(W/L)$ A would be constant. This property is used to model the threshold voltage, study its dependence on the device biasing, to investigate effects of localised charges, etc.

In this method, the tunnelling length at which the device current $I_{DS} = 10^{-7}$ A is calculated using either analytical or numerical methods [18–20]. Once this threshold tunnelling length is known, it is equated to the tunnelling length as a function of the bias obtained from the analytical model. The gate (or drain) bias at which the tunnelling length equals the threshold tunnelling length is the threshold voltage of the device.

In the constant current (or tunnelling length) method, the threshold voltage is extracted by equating the current with an arbitrarily defined (albeit practically useful) value. The values of the threshold voltage extracted using this method give little insight into the physical functioning of the TFET, as they have no physical basis. Due to this, the transconductance change (TC) method, which is widely used in MOSFETs, was proposed to extract the threshold voltage in TFETs.

6.3.3 Transconductance change (TC) method

The transconductance change method defines the threshold voltage for any non-linear device as the gate voltage at which the derivative of the transconductance dg_m/dV_{GS} is maximum [21]. This definition has been validated for MOSFETs and shown to give the same results as the physical definition of the threshold voltage. This method was subsequently extended [22] to extract the threshold voltage for TFETs.

Observing the transfer characteristics of the TFETs, we find that the dependence of the current on the gate bias changes from quasi-exponential to linear with increasing gate bias. Similarly, from the output characteristics of the TFET, we find that the dependence of the current on the drain bias changes from quasi-exponential to linear with increasing drain bias. This behaviour is the basis of threshold voltage extraction in the transconductance change method.

To calculate the gate threshold voltage, the transconductance g_m is obtained as a function of the gate bias V_{GS}. The gate threshold voltage is defined as the gate bias at which dg_m/dV_{GS} is maximum. Similarly, to calculate the drain threshold voltage, the conductance g_{ds} is obtained as a function of the drain bias V_{DS}. The threshold voltage is defined as the drain bias at which dg_{ds}/dV_{DS} is maximum.

The threshold voltages extracted using this method exhibit very different behaviour when compared to those extracted by the constant current method. The gate threshold voltage shows negligible roll-off with decreasing gate length (Figure 6.12), since the change in gate control from quasi-exponential to linear is independent of the gate length. Furthermore, the gate threshold voltage increases almost linearly with increasing drain voltage. The reason for this is that at a higher drain voltage, the channel potential would be pinned at a higher gate voltage, thereby causing a shift in gate control from quasi-exponential to linear at a higher gate voltage. Similarly, the drain threshold voltage also increases with increasing gate voltage, albeit in a non-linear manner (Figure 6.13).

References

[1] J. Konch, S. Mantl and J. Appenzeller, "Impact of the Dimensionality on the Performance of Tunneling FETs: Bulk Versus One-Dimensional Devices", *Solid State Electronics*, vol. 51, pp. 572–578, 2007.

[2] N. N. Mojumder and K. Roy, "Band-to-Band Tunneling Ballistic Nanowire FET: Circuit-Compatible Device Modeling and Design of Ultra-Low Power Digital Circuits and Memories", *IEEE Trans. Electron Devices*, vol. 56, pp. 2193–2201, October 2009.

[3] L. D. Michielis, L. Lattanzio, K. E. Moselund, H. Riel and A. M. Ionescu, "Tunneling and Occupancy Probabilities: How Do They Affect Tunnel-FET Behavior?", *IEEE Electron Device Lett.*, vol. 34, pp. 726–728, June 2013.

[4] M. Graef, T. Holtij, F. Hain, A. Kloes and B. Iniguez, "Two-Dimensional Physics-Based Modeling of Electrostatics and Band-to-Band Tunneling in Tunnel-FETs", *International Conference on Mixed Design of Integrated Circuits and Systems*, June 2013.

[5] E. Gnani, A. Gnudi, S. Reggiani and G. Baccarani, "Drain-Conductance Optimization in Nanowire TFETs by Means of a Physics-Based Analytical Model", *Solid State Electronics*, vol. 84, pp. 96–102, 2013.

[6] B. Bhushan, K. Nayak, and V. R. Rao, "DC Compact Model for SOI Tunnel Field-Effect Transistors", *IEEE Trans. Electron Devices*, vol. 59, no. 10, pp. 2635–2642, October 2012.

[7] E. O. Kane, "Zener Tunneling in Semiconductors," *Journal of Physics and Chemistry of Solids*, vol. 12, pp. 181–188, 1959.

[8] N. Cui, R. Liang, J. Wang and J. Xu, "Si-Based Hetero-Material-Gate Tunnel Field Effect Transistor: Analytical Model and Simulation," *12th IEEE International Conference on Nanotechnology (IEEE-NANO)*, 2012, pp. 1–5.

[9] N. Cui, R. Lianga, J. Wang and J. Xu, "Two-Dimensional Analytical Model of Hetero Strained Ge/Strained Si TFET", *International Silicon–Germanium Technology and Device Meeting (ISTDM)*, 2012, pp. 1–2.

[10] M. Lee and W. Choi, "Analytical Model of Single-Gate Silicon-on-Insulator Tunneling Field-Effect Transistors," *Solid State Electronics*, vol. 63, no. 1, pp. 110–114, 2011.

[11] L. Liu, D. Mohanta and S. Datta, "Scaling Length Theory of Double-Gate Interband Tunnel Field-Effect Transistors," *IEEE Trans. Electron Devices*, vol. 59, no. 4, pp. 902–908, April 2012.

[12] J. Wan, C. L. Royer, A. Zaslavsky and S. Cristoloveanu, "A Tunneling Field Effect Transistor Model Combining Interband Tunneling with Channel Transport", *J. Appl. Phys.*, vol. 110, no. 10, pp. 104503–104503-7, 2011.

[13] W. Vandenberghe, A. S. Verhulst, G. Groeseneken, B. Soree and W. Magnus, "Analytical Model for Point and Line Tunneling in a Tunnel Field-Effect Transistor," in *Proceedings of International Conference SISPAD*, 2008, pp. 137–140.

[14] A. S. Verhulst, D. Leoneli, R. Rooyackers and G. Groeseneken, "Drain Voltage Dependent Analytical Model of Tunnel Field-Effect Transistors," *J. Appl. Phys.*, vol. 110, no. 2, pp. 024510-1–024510-10, 2011.

[15] W. G. Vandenberghe, A. S. Verhulst, G. Groeseneken, B. Soree and W. Magnus, "Analytical Model for Tunnel Field-Effect Transistors," in *Proceedings of 14th IEEE Mediterranean Electrotechnology Conference*, May 2008, pp. 923–928.

[16] R. Vishnoi and M. Jagadesh Kumar, "An Accurate Compact Analytical Model for the Drain Current of a TFET from Sub-threshold to Strong Inversion," *IEEE Trans. Electron Devices*, vol. 62, no. 2, pp. 478–484, February 2015.

[17] R. Vishnoi and M. Jagadesh Kumar, "A Compact Analytical Model for the Drain Current of Gate All Around Nanowire Tunnel FET Accurate from Sub-threshold to ON-state," *IEEE Trans. Nanotechnology*, vol. 14, no. 2, pp. 358–362, March 2015.

[18] K. K. Bhuwalka, J. Schulze and I. Eisele, "Scaling the Vertical Tunnel FET with Tunnel Bandgap Modulation and Gate Work Function Engineering," *IEEE Trans. Electron Devices*, vol. 52, pp. 909–917, May 2005.

[19] N. Cui, R. Liang and J. Xu, "Heteromaterial Gate Tunnel Field Effect Transistor with Lateral Energy Band Profile Modulation", *Appl. Phys. Lett.*, vol. 98, pp. 142105–142105-3, 2011.

[20] R. Vishnoi and M. Jagadesh Kumar, "Two Dimensional Analytical Model for the Threshold Voltage of a Tunneling FET with Localized Charges," *IEEE Trans. Electron Devices*, vol. 61, no. 9, pp. 3054–3059, September 2014.

[21] R. Booth, M. White, H. Wong and T. Krutsick, "The Effect of Channel Implants on MOS transistor characterization", *IEEE Trans. Electron Devices*, vol. 34, pp. 2501–2509, 1987.

[22] K. Boucart and A.M. Ionescu, "A New Definition of Threshold Voltage in Tunnel FETs", *Solid State Electronics*, vol. 52, no. 9, pp. 1318–1323, 2008.

7

Device simulation using ATLAS

Technology computer aided design (TCAD) simulations are indispensable tools for any device modeller. They enable us to observe 1D, 2D and 3D variations in different physical quantities such as surface potential, electric field, band-to-band generation rate, electron/hole concentration, etc., that cannot be measured experimentally. Simulations can give us a clear insight into the functioning of the device, as the measurements here are not limited to the device terminals and can be done anywhere throughout the structure of the device. This can help a device modeller or a device designer to understand the effects of different parameters on the complete functioning of the device. However, simulations have their own limitations. One has to make sure that the simulation captures all the physical effects in the device accurately before putting them to any use.

Once the accuracy of the simulations is ensured, then simulations can be really useful for device modelling and for investigating new device designs. A simulation setup is inexpensive as compared to an experimental setup and also gives quick results. Therefore, one should formalise their ideas using simulations and then proceed towards fabricating their device.

In the present day, many commercial TCAD tools are available in the market. The leading ones are Silvaco ATLAS and Synopsys Sentaurus. Also, many organisations, universities and research centres have their own TCAD tools.

Tunnel Field-Effect Transistors (TFET): Modelling and Simulation, First Edition. Jagadesh Kumar Mamidala, Rajat Vishnoi and Pratyush Pandey.

In this chapter, we will primarily focus on TFET device simulations using Silvaco ATLAS [1]. At any point in the text, the reader can refer to the ATLAS user manual available at http://www.silvaco.com/.

In TFET simulations, accuracy is a major challenge. Most of the tunnelling models are solutions to the quantum mechanical phenomenon of tunnelling and use numerous approximations. Therefore, TFET simulation results can differ from the experimental results in orders of magnitude. Hence, it is always necessary to calibrate the simulations with experimental data before using the simulation results for formulating any new idea on TFETs.

In the rest of the chapter we focus on explaining how the ATLAS simulations works and the challenges faced in running the simulations.

7.1 Simulations using ATLAS

ATLAS is a physics-based device simulator that predicts the electrical behaviour of a defined semiconductor structure and gives insights into the physical mechanism operating inside the device. The structure of the semiconductor device is divided into 2D or 3D grids (or mesh), as shown in Figure 7.1, and a set of five basic physics-based semiconductor equations is solved in a coupled manner at each mesh point. These equations are Poisson's equation, the electron and hole continuity equation and the electron and hole current equations. Using these equations, the electron and hold density and their transport through the device structure is simulated. In other words, using numerical methods, a set of basic equations in

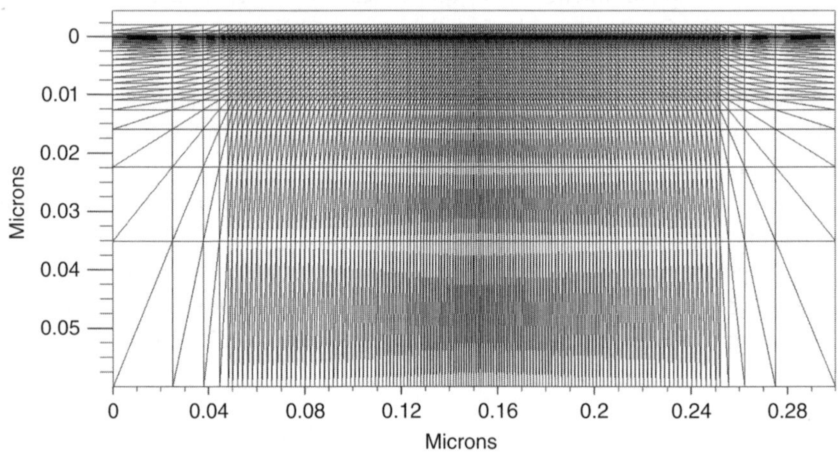

Figure 7.1 Example of a 2D mesh grid in ATLAS.

semiconductor physics and other user-specified physical models are solved in order to predict the operation of the device.

7.1.1 Inputs and outputs

Figure 7.2 shows the type of information flow in an ATLAS device simulation. It consists of two types of input files: a structure file, which defines the device structure and the mesh points, and a command file, which contains all the commands that ATLAS has to execute in order to perform the simulation. The simulation results are given in the form of three files. The first is the runtime output file, which tracks the progress of the simulation as it proceeds. The second is the log file, which saves all the terminal characteristics (currents and voltages) from the simulation. The third is the solution file, which saves the data of the 2D or 3D structure mesh points relating to the solution variables of the physical models solved for in the simulation at a given bias point.

The input file in ATLAS has a [.in] extension and contains both the structure and command inputs. An ATLAS input file is opened using DECKBUILD, an ATLAS interactive tool that provides an interactive runtime environment. Each line in the input file contains a statement. Each statement consists of a keyword (`<STATEMENT>`), which defines the statement and a set of parameters associated with the statement (`<PARAMETER>`). The general format of the statement is as follows:

```
<STATEMENT> <PARAMETER> <PARAMETER>=<VALUE>
```

For example:

```
DOPING UNIFORM N.TYPE CONCENTRATION=1.0e16 REGION=1
```

where `DOPING` is the keyword of the statement and `UNIFORM, N.TYPE, CONCENTRATION, REGION` are parameters. Any statement in ATLAS is case-invariant.

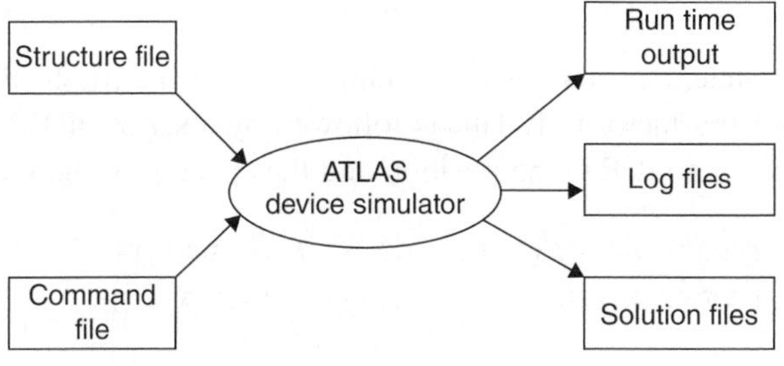

Figure 7.2 Flow of information in ATLAS.

There are five groups of statements that an ATLAS input file must contain in the correct order (as given below). These groups are:

1. **Structure specification.** This group of statements defines the structure of the model. This includes statements for defining the mesh, regions, electrodes, material, doping, etc.

2. **Materials and models specification.** This group of statements specifies the material parameters and physical models to be included in the simulations. This includes statements for defining material, models, contact properties, interface charges, etc.

3. **Numerical method specification.** This statement specifies the numerical methods to be used for solving the set of equations associated with the simulation.

4. **Solution specification.** This group of statements specifies the bias conditions for which the simulations have to be run. It includes statements that command the simulator to solve for the given bias conditions, log the solution and save it in a solution file.

5. **Result analysis.** This group of statements consists of commands to open and analyse the solution saved in a solution file.

Let us look briefly into each of the above five group of statements.

7.1.2 Structure specification

The first step of structure specification is defining the mesh. A mesh is a set of horizontal and vertical lines at the intersection of which the simulation equations are solved. The entire structure of the device has to be defined within the region covered by the mesh. A mesh is defined as follows:

```
MESH SPACE.MULT=<VALUE>
```

The above statement defines the scaling factor of the mesh. By default the value of the scaling factor is 1. This is followed by a series of X.MESH and Y. MESH statements, to define the position and the spacing of the mesh lines:

```
X.MESH  LOCATION=<VALUE>  SPACING=<VALUE>
Y.MESH  LOCATION=<VALUE>  SPACING=<VALUE>
```

The location and spacing between the mesh lines are specified in μm. ATLAS automatically inserts mesh lines between two specified X.MESH or Y.MESH locations according to the given spacing. An example of defining a 2D mesh is as follows:

```
x.mesh loc=0.00   spac=0.50
x.mesh loc=1.15   spac=0.02
x.mesh loc=1.5    spac=0.1
x.mesh loc=1.85   spac=0.02
x.mesh loc=3      spac=0.5
y.mesh loc=-0.017 spac=0.02
y.mesh loc=0.00   spac=0.005
y.mesh loc=0.1    spac=0.02
y.mesh loc=0.2    spac=0.01
y.mesh loc=0.6    spac=0.25
```

Note that any parameter term can be shortened. For example, `LOCATION` can be written as `loc` and `SPACING` as `spac`.

The second step is defining regions. Once the mesh is defined, each part of the mesh must be assigned a region and a material. This is done using the region statement:

```
REGION number=<integer> <position parameters>
<material_type>
```

For example:

```
region   num=1 y.max=0 oxide
region   num=2 y.min=0 y.max=0.2 silicon
region   num=3 y.min=0.2 oxide
```

This gives the regions shown in Figure 7.3.

While defining a mesh, it is always good practice to define the *x*-mesh and *y*-mesh locations at each line of interface between two regions. Otherwise, if a mesh line does not exist at the boundary of a region, then the nearest mesh line will be taken as the boundary. This may not give you the structure that you desired. Also, mesh lines should be made finer at the boundaries between two regions and in any other area where we expect a parameter to change rapidly, and the mesh can be made coarser in other areas. Making the mesh too fine can make the simulation runtime large. Hence, it is good practice to make the mesh as coarse as possible, without losing the capability of capturing the characteristics correctly.

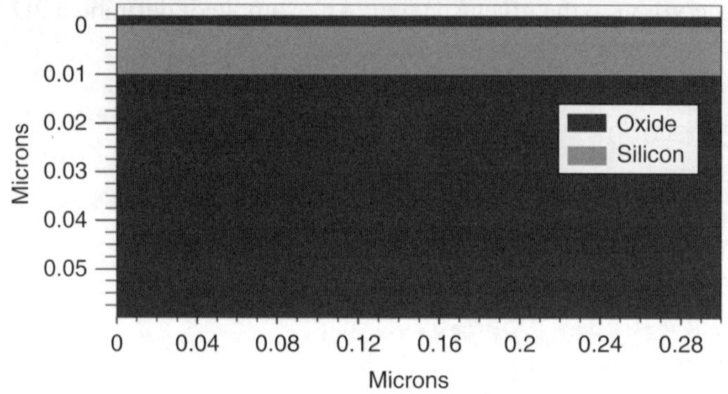

Figure 7.3 Example of regions of a device structure in ATLAS.

The third step is defining the electrodes. Using the electrode statement does this:

```
ELECTRODE NAME=<electrode name> <position_parameters>
```

It is required to define at least one electrode contacting a semiconductor material. For example:

```
electrode  name=gate x.min=1 x.max=2 y.min=-0.017 y.max=-0.017
electrode  name=source x.max=0.5 y.min=0 y.max=0
electrode  name=drain x.min=2.5 y.min=0 y.max=0
electrode  substrate
```

The substrate by default is taken at the highest position of *y* defined in the *y*-mesh statements. A very critical point here is to make sure that a mesh line is present along the boundary of the electrode, especially when we are defining a 1D electrode like the electrodes in the above example. Otherwise the entire electrode itself may get defined at the nearest mesh line, completely changing the effect of the electrode in the simulations.

The final part is to define the doping of the semiconductor regions. This is done with the doping statement:

```
DOPING <distribution_type> <dopant_type>
<position_parameters > .
```

For example:

```
doping   uniform conc=2e17 p.type reg=2
doping   gauss n.type conc=1e20 char=0.2 lat.char=0.05 reg=2 x.r=1.0
doping   gauss n.type conc=1e20 char=0.2 lat.char=0.05 reg=2 x.l=2.0
```

Doping profiles can be of different types and each type has its own set of parameters. In the above example, in the first statement, we define uniform doping in which we have to define two parameters: concentration (`conc=2e17`) and type (`p.type`). In the second and third statements, we define Gauss doping, which has the following parameters: type (`n.type`), peak concentration (`conc=1e20`), y-direction characteristic length (`char=0.2`) and lateral characteristic length (`lat.char=0.05`). The Gaussian doping profile is defined by the following expression:

$$N(y) = conc \times \exp\left(-\left(\frac{Y}{char}\right)^2\right) \tag{7.1}$$

7.1.3 Material parameters and model specification

After defining regions and materials in the structure specification section, in this section the first part is to specify the parameters of the materials (like bandgap, mobility, permittivity, etc.) defined in the structure specification section. This is done using the material statement. For example:

```
MATERIAL MATERIAL=Silicon EG300=1.12 MUN=1100
```

This sets the bandgap (`EG300`) and low field mobility (`MUN`) of silicon regions defined in the structure. If this part is skipped, then default values of material parameters are taken.

The second part is to define the contacts. This is done using the contact statement. For each electrode defined in the structure specification section, there is a contact statement where the work function or the material of the electrode is defined. For example:

```
contact name=gate n.poly;
CONTACT NAME=gate WORKFUNCTION=4.8;
```

Here we have two choices. We can either specify the material of the electrode, which will include all the default parameter values associated with the specified material in the simulation, or we can specify specific properties of the electrode material like work function, resistance, capacitance, inductance, etc. If no work function or material type is specified then the electrode is assumed to be an ohmic contact.

The third part is defining the physical models to be included in the simulations. Models of basic semiconductor physics (like Boltzmann carrier statistics, drift–diffusion, etc.) are defined by default and need not be defined in this section.

The physical models can be distributed into five categories: mobility, recombination, carrier statistics, impact ionisation and tunnelling. All models are defined using the model statement. Only the impact ionisation models are defined using the impact statement. For example:

```
MODELS CONMOB FLDMOB SRH FERMIDIRAC;
IMPACT SELB;
```

The above statement specifies standard concentration-dependent mobility, parallel field mobility, Shockley–Read–Hall recombination with fixed carrier lifetimes and Fermi–Dirac statistics. The Selberherr impact ionisation model is also specified here.

7.1.4 Numerical method specification

After defining the structure and the models, we need to define the numerical methods that ATLAS has to use in order to solve for the specified physical model equations and place on to the defined structure. This is done using the model statement. For example:

```
METHOD GUMMEL NEWTON;
```

This statement specifies that the Gummel method will be used to solve the physical model equations followed by the Newton method, in case the solution does not converge using the Gummel method.

7.1.5 Solution specification

After setting up the structure, models and methods of solution, we have to now obtain the solutions or, in other words, run the simulation. This is done by using the solve statement followed by specifying the bias. The syntax of this statement is as follows:

```
SOLVE VGATE=1.0;
```

There are two important rules here. First, any electrode for which the bias voltage is not defined in the given solve statement needs to have its value specified in the previous solve statement taken. Second, if the bias voltage of an electrode is never defined in a solve statement then its value is taken to be zero. For most of the analysis, sweeping one or more bias voltages is required. This is done as follows:

```
solve vgate=0.1 vstep=0.1 name=gate vfinal=1.5
```

This sweeps the voltage at the gate electrode from 0.1 to 1.5 V in steps of 0.1 V. A typical example of a solution specification is:

```
solve    init
solve    prev
solve    vgate=-0.2
solve    vdrain=0.05
solve    vdrain=0.1
# ramp gate voltage
solve   vgate=0.1 vstep=0.1 name=gate vfinal=1.5
```

As shown above using the symbol #, we can specify a comment.

7.2 Analysis of simulation results

As shown in Figure 7.2, there are three types of outputs from an ATLAS simulation. The first type is the runtime output, which is displayed in a window at the bottom of DECKBUILD (Figure 7.4). This shows the status of the simulation as each statement of the input file is executed. The second type of output is the log file. To save information into a log file, we use the log statement. Its syntax is as follows:

```
LOG OUTF=<FILENAME>
```

To save the data into a log file, one has to open a log file using the log statement given above. Terminal characteristics of all solve statements following the log statement are stored in the log file. To stop the log file from storing more data, we have to close the log file using the logoff statement.

For example:

```
log     outf=mosfet.log
solve   vgate=0.1 vstep=0.1 name=gate vfinal=1.5
log off
```

The above statements store the terminal characteristics of the gate voltage sweep into the log file named mosfet.log and then close the log file.

The third type of output is the solution file (or the structure file). This is generated using the save statement. Its syntax is

```
save    outf=<filename>.str
```

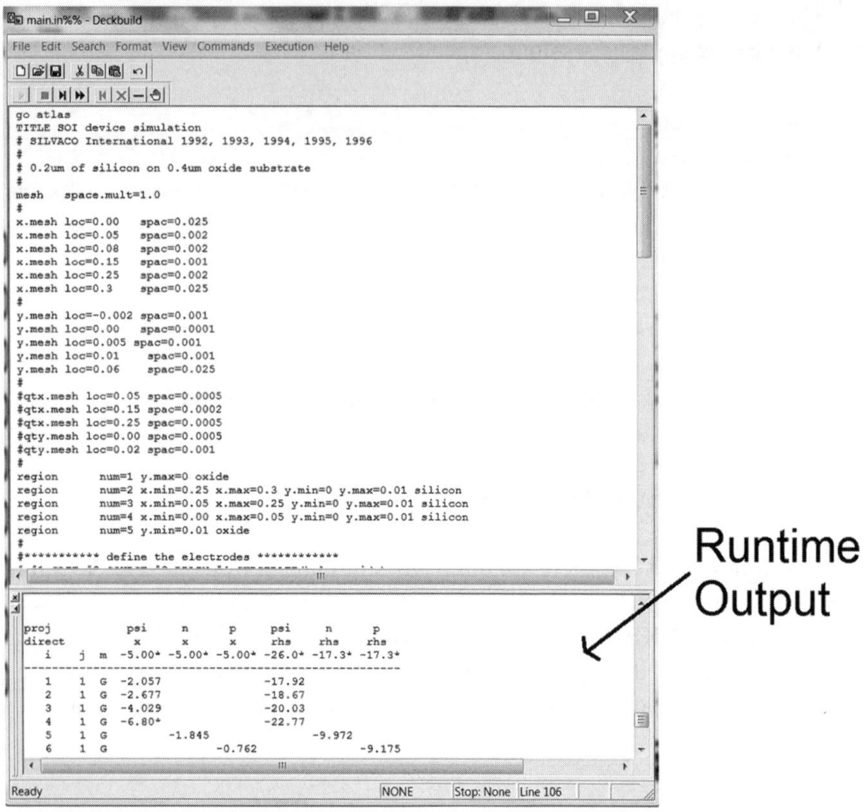

Figure 7.4 The DECKBUILD window.

The above save statement saves the solution data corresponding to each mesh point defined in the structure at the last specified bias point into the structure (.str) file. For example:

```
save      outf=mosfet.str
```

The next part is viewing the log and the structure files stored in the simulation. This is done by opening the files in TONYPLOT, which is an ATLAS interactive tool like DECKBUILD and is used for plotting data. The following statement is used in the input file or separately in the command terminal to open the log and structure files using TONYPLOT:

```
tonyplot <filename>.log or tonyplot <filename>.str
```

Figure 7.5 shows a log file and Figure 7.6 shows a structure file opened in TONYPLOT, respectively. The log file opens as a 2D plot, where we can choose the quantities to display on the *x*- and *y*-axes using the graphical user interface (GUI) of TONYPLOT. A structure file opens as a 2D schematic of the structure of the device showing different regions and electrodes of the device (such as source, drain, gate oxide, etc.). To view the spatial distribution of different physical quantities (potential, electric field, doping, etc.) we have to make a horizontal

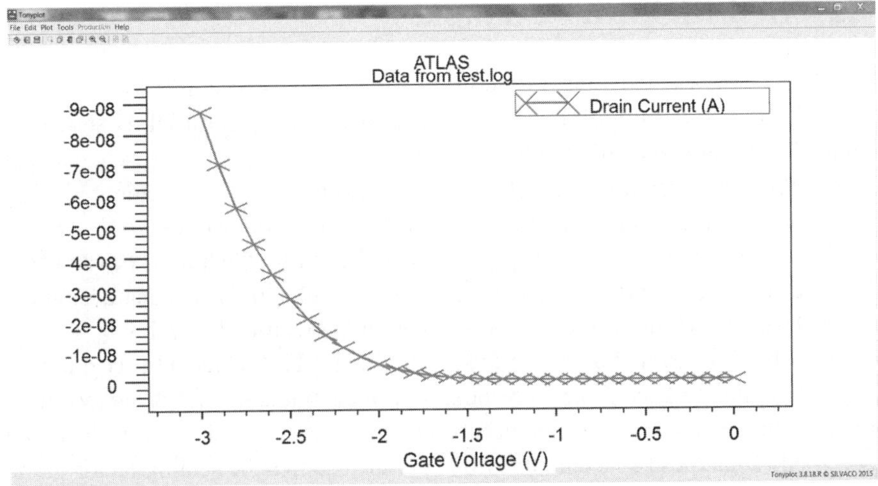

Figure 7.5 Example of a log file in ATLAS.

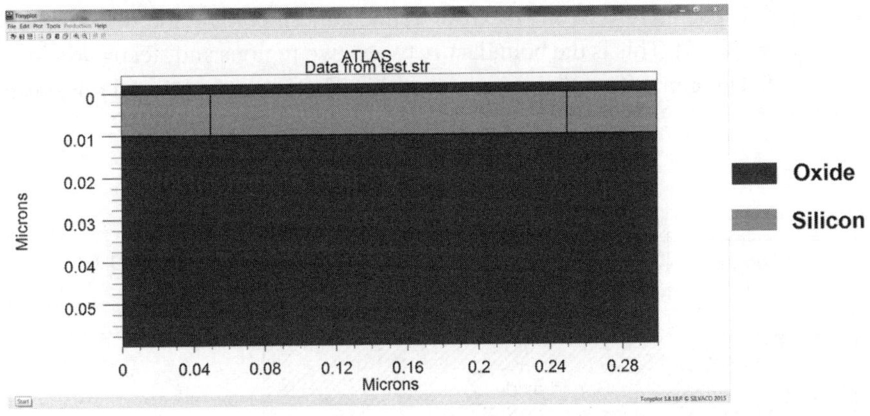

Figure 7.6 Example of a structure file in ATLAS.

or vertical cutline using the cutline option. This opens a 2D plot where we can choose to plot different physical quantities along the cutline. In the 2D schematic of the structure itself, we can also see the mesh lines and contours of different physical quantities.

After understanding the basics of ATLAS simulation, in the following section we will illustrate how to run an ATLAS simulation using an example of a basic device structure.

7.3 SOI MOSFET example

In this section, we will illustrate how to run an ATLAS simulation, with the example of a SOI MOSFET. The structure of the SOI MOSFET is shown in Figure 7.7 To simulate this structure the following parameters are chosen: channel length (L) = 1 μm, length of source and drain regions = 1 μm, source/drain doping (Gaussian profile) peak concentration = 10^{20}/cm^3, body doping (N_A) = 2 × 10^{17}/cm^3, gate oxide thickness (T_{ox}) = 17 nm, silicon film thickness (T_{Si}) and buried oxide thickness = 400 nm. The gate is of n-type poly-silicon and has a length = 1 μm. The ATLAS input file for the simulation is given in Table 7.1.

In Table 7.1, `go atlas` (Line 1) specifies that ATLAS has to be used for the simulation and `TITLE` (Line 2) is the title of the simulation and is always mentioned at the second line of the input file, but this is optional and may not always be specified. Then mesh space.mult = 1.0 (Line 3) specifies the scaling factor of the mesh which is defined equal to 1. The statement `x.mesh` (Lines 4 to 8) defines the x-mesh. Here we start with defining a mesh point at $x = 0$, which is the starting point of the structure. Since $x = 0$ is the beginning of the source region, not much of a spatial variation in physical quantities is expected here and hence we define a coarse mesh spacing of 0.5 μm. Then we define a mesh point at the source–body junction (x = 1.15). This is the boundary between two regions and steep variations are expected here in different physical quantities. Therefore we define a fine mesh

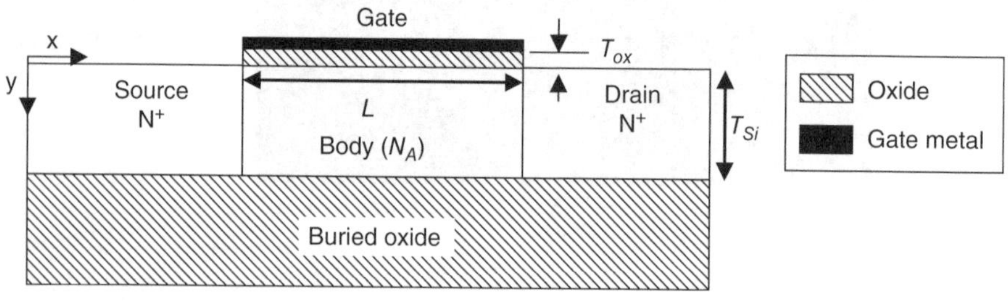

Figure 7.7 Schematic of the SOI MOSFET under study.

Table 7.1 Example code in ATLAS for SOI MOSFET simulation

```
Line1:    go atlas
Line2:    TITLE SOI device simulation
# SILVACO International 1992, 1993, 1994, 1995, 1996
#
# 0.2um of silicon on 0.4um oxide substrate
#
Line3:    mesh  space.mult=1.0
#
Line4:          x.mesh loc=0.00  spac=0.50
Line5:          x.mesh loc=1.15  spac=0.02
Line6:          x.mesh loc=1.5   spac=0.1
Line7:          x.mesh loc=1.85  spac=0.02
Line8:          x.mesh loc=3     spac=0.5
#
Line9:          y.mesh loc=-0.017 spac=0.02
Line10:         y.mesh loc=0.00   spac=0.005
Line11:         y.mesh loc=0.1    spac=0.02
Line12:         y.mesh loc=0.2    spac=0.01
Line13:         y.mesh loc=0.6    spac=0.25
#
Line14:        region   num=1 y.max=0 oxide
Line15:        region   num=2 y.min=0 y.max=0.2 silicon
Line16:        region   num=3 y.min=0.2 oxide
#
#*********** define the electrodes ***********
# #1-GATE #2-SOURCE #3-DRAIN #4-SUBSTRATE(below oxide)
#
Line17: electrode name=gate  x.min=1 x.max=2 y.min=
  -0.017 y.max=-0.017
Line18: electrode name=source x.max=0.5 y.min=0 y.max=0
Line19: electrode name=drain x.min=2.5 y.min=0 y.max=0
Line20: electrode substrate
#
#*********** define the doping concentrations *****
#
Line21:  doping  uniform conc=2e17 p.type reg=2
Line22:  doping  gauss n.type conc=1e20 char=0.2 lat.
  char=0.05 reg=2 x.r=1.0
Line23:  doping  gauss n.type conc=1e20 char=0.2 lat.
  char=0.05 reg=2 x.l=2.0
#
```

(continued overleaf)

Table 7.1 (*continued*)

```
# set interface charge separately on front and back oxide
  interfaces
Line24:        interf     qf=3e10 y.max=0.1
Line25:        interf     qf=1e11 y.min=0.1
#
# set workfunction of gate
Line26:        contact    name=gate n.poly
#
# select models
Line27:        models     conmob srh auger bgn fldmob print
#
Line28:        solve init
#
# do IDVG characteristic
#
Line29:        method     newton   trap
Line30:        solve      vdrain=0
Line31:        solve      vdrain=0.05
Line32:        solve      vdrain=0.1
#
# ramp gate voltage
Line33:        log        outf=soiex.log
Line34:        solve      vgate=0.1 vstep=0.1 name=gate vfinal=1.5
Line35:    save       outf=test.str
#
#
Line36:        quit
```

spacing of 0.02 μm here. Now, in the centre of the channel, we expect a gradual spatial variation in physical quantities and hence we can afford a coarser mesh here. Then we define a mesh point at $x = 1.5$ with a spacing of 0.1 μm. Similar to the source–body junction we define a finer mesh point at the drain body junction, that is $x = 1.85$, with a spacing of 0.01 μm. Finally, we define the end point of the device at $x = 3$ with a spacing of 0.5 μm, similar to that at the starting point.

The statement `y.mesh` (Lines 9 to 13) is used to specify the mesh lines of the structure in the y-direction. It is common practice to place the silicon–dielectric interface at $y = 0$. Since the gate oxide thickness in this example is 17 nm, we start by defining our starting mesh point at $y = -0.017$. This is the position of the gate.

Note that although the gate is shown to have a finite thickness in Figure 7.7, in an input file we can define the gate with zero thickness and a finite length. Between $y = -0.017$ and $y = 0$, we need a fine meshing (e.g. 0.02 μm) because these mesh lines are inside the thin dielectric where we expect to observe strong variation in physical quantities as we move along the y-direction. Now we define the next mesh point at $y = 0$, that is the silicon–dielectric interface with a very fine mesh spacing of 0.005 μm. These mesh lines would be inside the channel and hence we require a very fine mesh in this region. As we move away from the silicon–silicon dioxide interface in the y-direction, we do not need such a fine mesh away from the inversion layer. Therefore, we define the next mesh point at $y = 0.1$ with a spacing of 0.02 μm. Now at the silicon–buried oxide interface, that is at $y = 0.2$, we define the next mesh point with a spacing of 0.01 μm. Next we define the end point of the mesh at $y = 0.6$, with a coarse spacing of 0.25 μm. These mesh lines are inside the buried oxide, where we expect a very gradual spatial variation in the physical quantities. Figure 7.8 shows the mesh grid of the structure generated after running the mesh statements of the above file.

Having defined the mesh of the structure, we now need to define the different regions of the structure. The statement `region` (Lines 13 to 15) is used for this purpose. The device has three regions. Each region is identified by a number (e.g. `num=1`). Number 1 in the region statement is the gate oxide region, which starts form the first y-mesh line and ends at $y = 0$ and extends throughout the device in the x-direction, that is from $x = 0$ μm to $x = 0.3$ μm. Number 2 is the silicon region, which starts from $y = 0$ to $y = 0.02$ and extends throughout the device in the

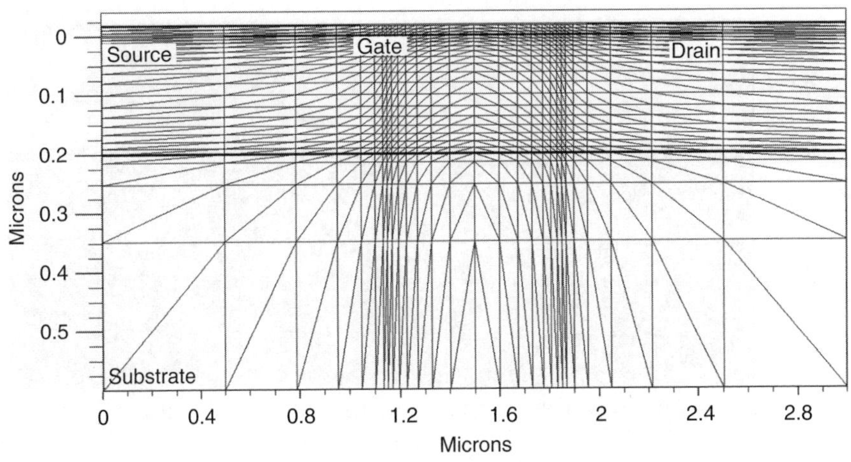

Figure 7.8 Mesh grid of the SOI MOSFET example.

x-direction. Number 3 is the buried oxide region, which starts at $y = 0.02$ and ends at the last y-mesh point and extends throughout the device in the x-direction.

The statement `electrode` (Lines 16 to 19) is used to define different electrodes in the device structure. First we define the gate electrode, which is a 1D electrode at the top of the structure, that is at $y = -0.017$, and extends from $x = 1\ \mu m$ to $x = 2\ \mu m$. Next, we define the source electrode, which is a 1D electrode at $y = 0$ and extends from $x = 0\ \mu m$ to $x = 0.5\ \mu m$. The source electrode is, generally, separated by the gate electrode using a sidewall oxide. Therefore, it is preferable to define the source electrode not covering the entire source region. This is followed by the drain electrode, which is a 1D electrode at $y = 0$ and extends from $x = 0.25\ \mu m$ to $x = 3\ \mu m$. Lastly, we define the substrate. If the coordinates of the substrate electrode are not defined, the substrate electrode is taken to be at the maximum y-mesh point extending all over the device structure in the x-direction. Figure 7.9 shows the electrodes of our device.

The statement `doping` (Lines 20 to 22) is used to define the type and the profile of doping in the silicon region of the device. We begin with defining a uniform doping of $2 \times 10^{17}\ cm^{-3}$ (p-type), in the entire silicon region of the device (Line 20). Next (Lines 21 and 22) we define a Gaussian doping profile in the source and drain regions of the device of n-type with peak concentration = 1e20/cm^3, y-direction characteristic length = 0.2 μm and lateral characteristic length = 0.05 μm. The statement `interf` (Lines 23 and 24) is used to define the interface charges at the gate oxide–silicon interface and buried oxide–silicon interface. The statement `contact` defines the gate electrode to be n-type polysilicon (Line 25). This defines the properties (work function, resistivity, inductance, capacitance, etc.) of

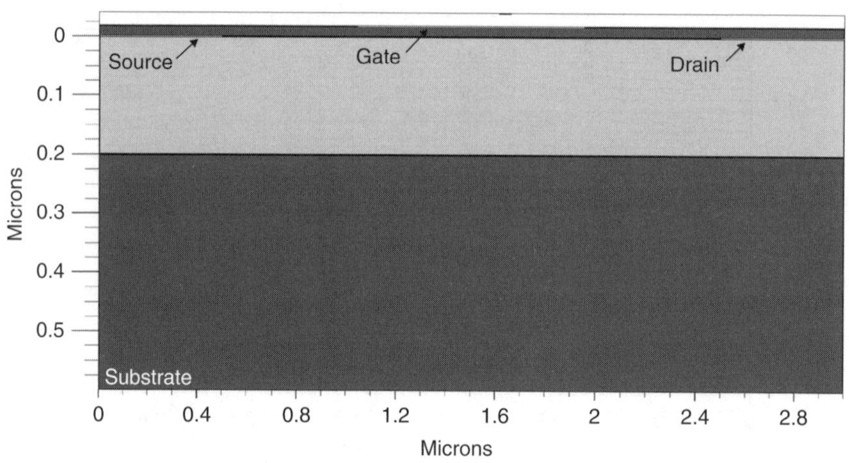

Figure 7.9 Electrodes of the SOI MOSFET example.

the gate material. The `model` statement (Line 26) describes the types of model to be used in the simulation. In this simulation, we include the models for concentration-dependent mobility (`conmob`), srh recombination (`srh`), auger recombination (`auger`), bandgap narrowing (`bgn`) and lateral electric field dependent mobility (`fldmob`). The parameter print prints the status of all models (i.e. coefficients and constants) in the runtime output. The statement `solve init` (Line 27) runs the simulation for finding the initial solution, that is with all terminals at zero bias. In the `method` statement (Line 28) we define that Newton and trap methods will be used for solving the physics-based semiconductor equations and the specified model (in Line 26) equations. Now we run the simulation for finding the I_D–V_{GS} characteristics of the device in the linear region. Using the `solve` statement (Lines 29 to 31), we solve the simulation for the desired bias points. Using the `log` statement (Line 32), we open a log file named soiex.log, which is followed by ramping the gate voltage from 0.1 to 1.5 V in steps of 0.1 V (Line 33), and the terminal characteristics of this voltage sweep get stored in the log file soiex.log. Finally, we save the structure file of the device using the save statement (Line 35).

Figure 7.10 shows the structure file (`soiex.str`) of the SOI MOSFET opened in TONYPLOT and the doping contours and different regions of the device. Figure 7.11 shows the log file (`soiex.log`) opened in TONYPLOT. It displays the transfer characteristics (I_D versus V_{GS}) of the device, corresponding to the gate voltage sweep in Line 34 of Table 7.1 at a drain voltage of 0.1 V.

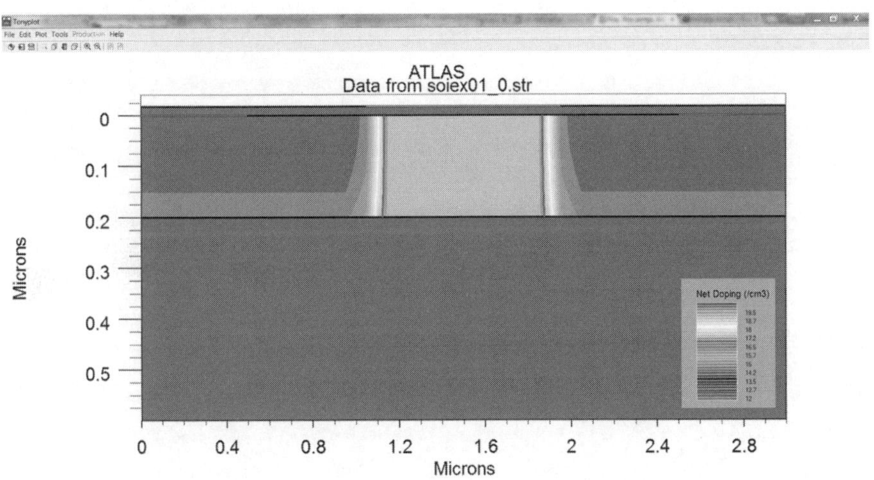

Figure 7.10 The structure file of the SOI MOSFET example showing the different regions and the doping profile (in log scale) in the device.

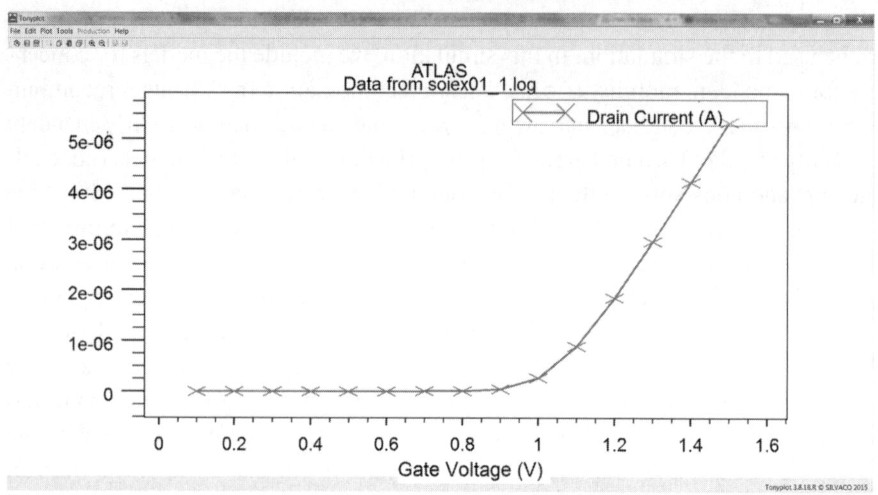

Figure 7.11 The log file of the SOI MOSFET example showing the drain current (I_D) versus the gate voltage (V_{GS}) characteristics.

Reference

[1] *ATLAS Device Simulation Software*, Silvaco Int., Santa Clara, CA, USA, 2015.

8

Simulation of TFETs

In the previous chapter, we started with basic device simulation using ATLAS [1]. We will now move towards the simulation of TFETs for 2D as well as 3D device structures. In the simulation of TFETs, we use different tunnelling models that require calibration with experimental data. In this chapter, we will discuss the simulation of a 2D SOI TFET followed by the simulation of a 3D gate all around nanowire TFET.

8.1 SOI TFET

A TFET has a structure similar to that of a MOSFET with only a difference in the type of source doping. Most of the tunnelling in a TFET occurs at the source–channel junction and hence this region of the device becomes important in simulation for predicting the results accurately. The two key points in the simulation of a TFET are defining an appropriate mesh and choosing the right tunnelling model parameters.

In this section, we will describe the simulation of an SOI TFET. Figure 8.1 shows the schematic of an SOI TFET. It has a channel length (L) of 200 nm and a gate oxide thickness (T_{ox}) of 2 nm. The thickness of the silicon film (T_{Si}) is 10 nm. The source/drain regions have a doping of 10^{21}/cm^3 and a length of 50 nm. This is a p-channel TFET and hence the source doping is of n-type and the drain doping is of p-type. The buried oxide thickness is 180 nm. The gate metal

Tunnel Field-Effect Transistors (TFET): Modelling and Simulation, First Edition. Jagadesh Kumar Mamidala, Rajat Vishnoi and Pratyush Pandey.

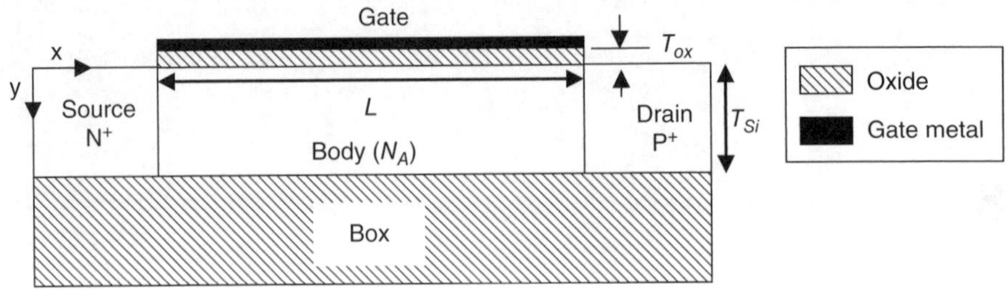

Figure 8.1 Schematic of the SOI TFET under study.

wok function is 4.8 eV in order to enhance band-to-band tunnelling at the source channel junction. The input file for the TFET simulation is shown in Table 8.1.

Section 1 of the file given in Table 8.1 defines the mesh of the structure. In a TFET, most of the tunnelling occurs at the source–channel junction and hence it is necessary to define a finer mesh at this junction. Due to a low ON-state current, there is less variation in physical quantities in the rest of the channel and hence we can afford to have a coarser mesh in the channel. Drain side tunnelling is an important phenomenon in the TFET and hence we should define a finer mesh at the drain–channel junction. In the *y*-direction, the electrostatics is mostly like a MOSET unless we are defining a vertical tunnelling structure and hence *y*-meshing is similar to that in the SOI MOSFET example described previously (Section 7.3). Sections 2 to 5 of the file are similar to the SOI MOSFET except that the source doping is n-type in the SOI TFET. Section 6 is the most important part of the simulation. Here we have to choose an appropriate tunnelling model and the correct values of the parameters of the tunnelling model. Although all the default parameter values in ATLAS are defined for silicon-based devices, using these default parameters for tunnelling can give an orders of magnitude difference between the simulations and the experimental results. Hence, it is very important to calibrate the simulation results with the characteristics of an experimental device before using the simulation for further study of the TFET. In this example, we have used Kane's model for the band-to-band tunnelling model, which is a local model used for tunnelling. The tunnelling parameters given in Section 6 (`a.btbt`, `b.btbt` and `bbt.gamma`) of the file are for Kane's model equation:

$$G_{BBT} = \frac{D \times BBT.A\_KANE}{E_g^{0.5}} F^{BBT.GAMMA} \exp\left(-BBT.B\_KANE \frac{E_g^{1.5}}{F} \right) \tag{8.1}$$

where G_{BBT} is the tunnelling generation rate, F is the local electric field, E_g is material bandgap, D is the statistical factor and *BBT.A_KANE* (`a.btbt`), *BBT.GAMMA* (`bbt.gamma`) and *BBT.B_KANE* (`b.btbt`) are the tunnelling

Table 8.1 Example code in ATLAS for an SOI TFET simulation

```
go atlas
TITLE SOI TFET device simulation
# SILVACO International 1992, 1993, 1994, 1995, 1996
#
# 10 nm of silicon on 180 nm oxide substrate
#
#Section 1: Defining the Mesh
mesh  space.mult=1.0
#
x.mesh loc=0.00   spac=0.025
x.mesh loc=0.05   spac=0.0002
x.mesh loc=0.08   spac=0.0002
x.mesh loc=0.15   spac=0.001
x.mesh loc=0.25   spac=0.002
x.mesh loc=0.3    spac=0.025
#
y.mesh loc=-0.002 spac=0.001
y.mesh loc=0.00   spac=0.0001
y.mesh loc=0.005  spac=0.001
y.mesh loc=0.01   spac=0.001
y.mesh loc=0.06   spac=0.025
#
#Section 2: Defining the regions
region  num=1 y.max=0   oxide
region   num=2 x.min=0.25 x.max=0.3 y.min=0 y.max=0.01
  silicon
region   num=3 x.min=0.05 x.max=0.25 y.min=0 y.max=0.01
  silicon
egion    num=4 x.min=0.00 x.max=0.05 y.min=0 y.max=0.01
  silicon
region  num=5 y.min=0.01 oxide
#
#Section 3: Defining the electrodes
# #1-GATE #2-SOURCE #3-DRAIN #4-SUBSTRATE (below the
  buried oxide)
#
electrode    name=gate  x.min=0.05  x.max=0.251 y.min=
  -0.002 y.max=-0.002
electrode    name=source x.max=0.05  y.min=0   y.max=0
electrode    name=drain  x.min=0.251 y.min=0   y.max=0
electrode substrate
#
```

(*continued overleaf*)

Table 8.1 (*continued*)

```
#Section 4: Defining the doping concentrations
#
doping    uniform conc=1e21  p.type  reg=2
doping    uniform conc=1e21  n.type  reg=4
doping    uniform conc=6.5e15 n.type  reg=3
#
#Section 5: Defining material properties
contact    name=gate workfunction=4.8
#
# Section 6: Defining the models
models  conmob srh auger bgn fldmob print bbt.kane a.
  btbt=4e19 b.btbt=41e6 bbt.gamma=2
#
#Section 7: Solving the simulation
solve init
#
method    gummel  newton
#
solve    vsource=0.0
solve    vdrain=0.0
solve    vdrain=0
solve    vdrain=-0.5
solve    vdrain=-1
#for output characteristics uncomment lines below
#solve    vgate=-0
#solve    vgate=-1
#solve    vgate=-1.5
#solve    vgate=-2
#
# ramp the gate voltage
log      outf=test.log master
solve    vgate=0 vstep=-0.1 name=gate vfinal=-3
#for output characteristics uncomment line below
# solve     vdrain=0 vstep=-0.1 name=drain vfinal=-3
output val.band con.band u.bbt charge e.lines band.param
save     outf=test.str
#Section 8: Reading the results
tonyplot   test.log
tonyplot   test.str
#
quit
```

parameters. These parameters are calibrated against experimental results [2, Figure 6(a)], as shown in Figure 8.2 [3]. The simulator calculates the tunnelling rate (G_{BBT}) and integrates it in the entire silicon film to find the drain current. As can be seen from Equation (8.1), tunnelling is an exponential function of the electric field and hence the tunnelling rate is maximum at the source–channel junction, where the electric field is maximum. At all other locations the tunnelling generation rate is negligible, as can be seen in Figure 8.3. According to Kane's paper [4] the tunnelling parameter *BBT.GAMMA* is equal to 2 for direct bandgap materials like Ge and 2.5 for indirect bandgap materials like Si. However, *BBT.GAMMA* can be fine-tuned between 2 and 2.5. The statistical factor *D* is by default taken to be 1, but for a TFET, this may give non-zero current at zero drain voltage.

Hence, while simulating output characteristics, one should include either `bbt.hurkx`, `bbt.dehurkx` or `bbt.djhurkx` models for finding the value of *D*. All the other models defined here are similar to that in the SOI MOSFET example described in Section 7.3. In Section 7 of Table 8.1, we specify both the `gummel` and the `newton` methods, which are the numerical methods for solving the simulation equations. This specification starts the process of finding the solution with the Gummel method, which solves the equations in a fully decoupled way and makes an initial guess. This is followed by the use of the Newton method, which solves the equations in a fully coupled way to find the final solution. This ensures better and faster convergence of the simulation. Also, in the output statement, we specify statements for including valance band energy,

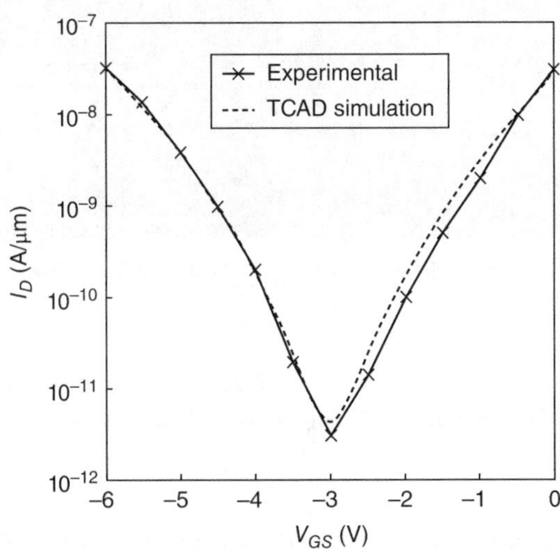

Figure 8.2 Reproduction of experimental results [2, Figure 6(a)] to calibrate the tunnelling parameters [3]. Source: Reproduced with permission of IEEE.

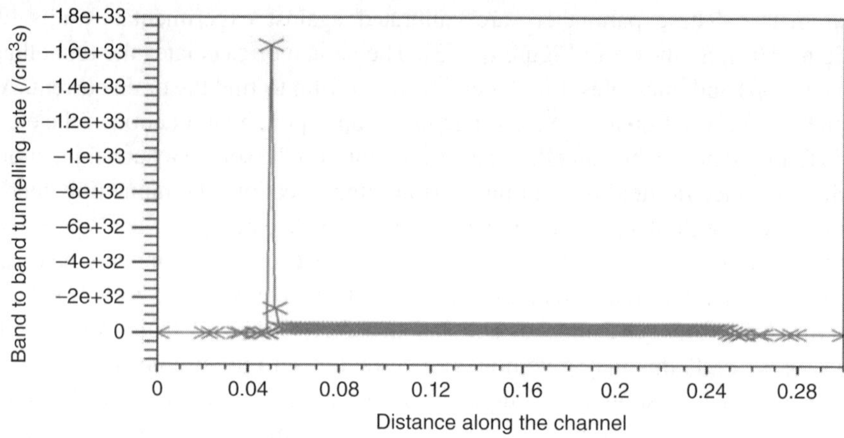

Figure 8.3 Tunnelling generation rate at the silicon–oxide interface (i.e. at y = 0) along the channel (i.e. x-direction).

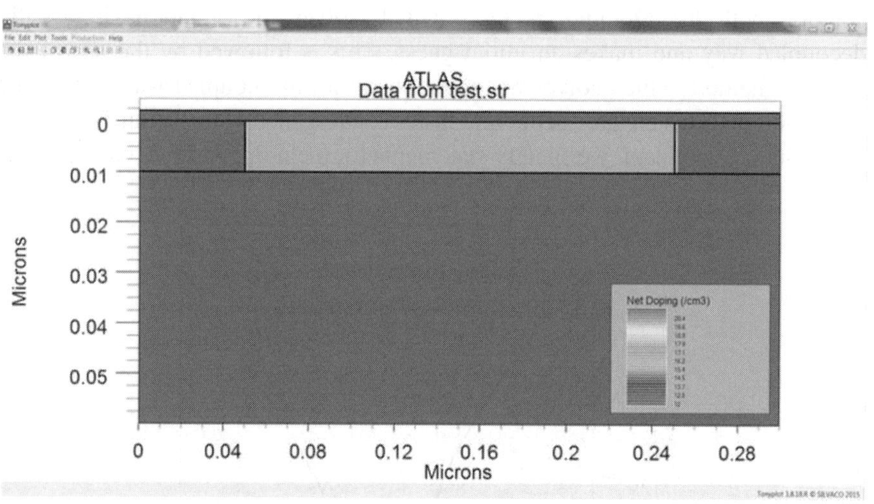

Figure 8.4 The structure file of the SOI TFET example showing the different regions and the doping profile (in log scale) in the device.

conduction band energy, tunnelling generation rate, net charge, electric field lines and band parameters in the standard structure (`test.str`) file.

Figure 8.4 shows the structure file (`test.str`) open in TONYPLOT. It displays the doping profile and different regions of the device. Figure 8.5 show a

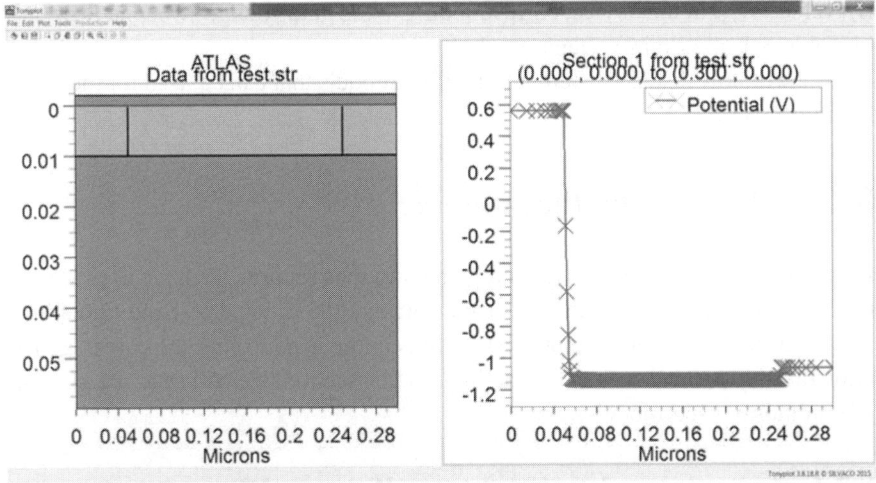

Figure 8.5 The structure file of the SOI TFET showing the potential plotted along a horizontal cutline made at the surface of the device (i.e. at y = 0).

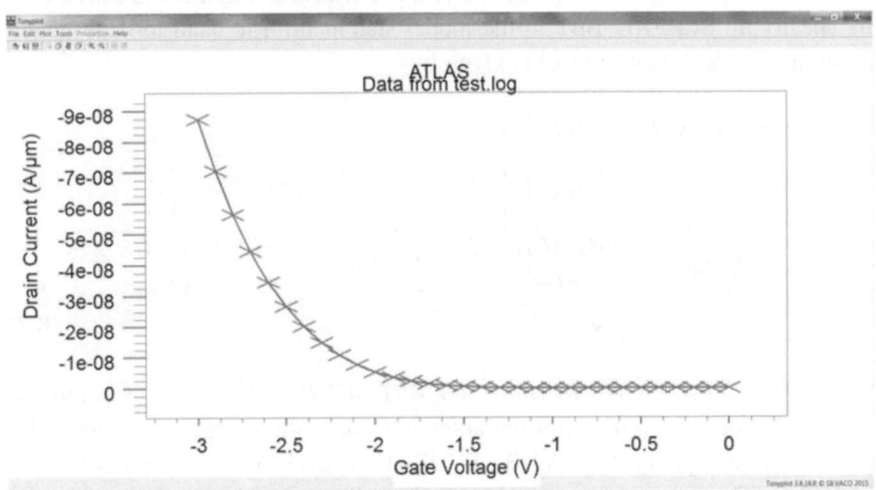

Figure 8.6 The log file of the SOI TFET example showing the drain current (I_D) versus gate voltage (V_{GS}) characteristics.

horizontal cutline made at the surface in the structure file (`test.str`) and the potential profile plotted along the cutline. Figure 8.6 shows the log file (`test.log`) open in TONYPLOT. It displays the transfer characteristics

(I_D versus V_{GS}) of the device, corresponding to the gate voltage sweep in Section 7 of Table 8.1 at a drain voltage of –1 V.

We will now discuss different tunnelling models that can be used for the simulation of a TFET.

8.2 Other tunnelling models

In the SOI TFET example described in the previous section, we have used Kane's model to find the generation rate of the carriers due to band-to-band tunnelling. However, there are some other models that can be used to find the carrier generation rate due to band-to-band tunnelling. In this section, we will describe a couple of commonly used models.

8.2.1 Schenk band-to-band tunnelling model

A model that includes a comprehensive study of phonon-assisted tunnelling or indirect tunnelling, is the Schenk band-to-band tunnelling model. This model neglects direct band-to-band tunnelling and hence is only good for indirect tunnelling materials such as silicon. It is also a local model like Kane's model and can be used by specifying `SCHENK.BBT` in the model statement. The band-to-band generation rate in the Schenk model is given by

$$G_{BBT} = A.BBT.SCHENK \times F^{7/2} \times S \times \left(\frac{(A^{\mp})^{-3/2} \exp\left(\frac{A^{\mp}}{F}\right)}{\exp\frac{(HW.BBT.SCHENK)}{kT} - 1} + \frac{(A^{\pm})^{-3/2} \exp\left(\frac{A^{\pm}}{F}\right)}{1 - \exp\frac{(-HW.BBT.SCHENK)}{kT}} \right) \tag{8.2}$$

where $A^{\pm} = B.BBT.SCHENK\,(\hbar\omega \pm HW.BBT.SCHENK)^{3/2}$, S is the statistical factor dependent on carrier concentration and $\hbar\omega$ is the energy of phonon. This model has three parameters, A.BBT.SCHENK, B.BBT.SCHENK and HW.BBT.SCHENK. These three parameters have to be calibrated with experimental results. As compared to Kane's model, this model is more accurate for silicon as it is an indirect bandgap semiconductor, but cannot be used for direct bandgap materials [5].

8.2.2 Non-local band-to-band tunnelling

Both Kane's and Schenk's models are local tunnelling models, which find the tunnelling generation rate at a given point based on the electric field at that point.

A non-local band-to-band tunnelling model (`bbt.nonlocal`), on the other hand, finds the tunnelling generation rate across a tunnelling length and incorporates the change in the electric field along the tunnelling length. This model is more accurate for reverse biased tunnelling junctions with high doping. To use this model, we need to define a special mesh called the qt mesh in the tunnelling region. Also, the tunnelling direction has to be defined separately using `qtunnel.dir` in the model statement. This model finds the tunnelling rate in parallel slices (which are defined by the qt mesh) along the tunnelling direction. The qt mesh interpolates data between underlying device mesh points (defined by the initial mesh statements) and performs tunnelling calculations. The qt mesh should contain only a single p–n junction and should cover it entirely. To begin with, a coarser qt mesh can be defined, which can then be made finer till the point where no change in simulation results is observed. The following describes the qt meshing for the SOI TFET given in Figure 8.1:

```
qtx.mesh loc=0.04 spac=0.0001
qtx.mesh loc=0.1 spac=0.0002
qty.mesh loc=0.000 spac=0.00001
qty.mesh loc=0.002 spac=0.00005
```

There are other ways for defining the qt mesh as well, which can be found in the ATLAS manual. The non-local band-to-band tunnelling model integrates the tunnelling of carriers across the bandgap at all the energies that lie between the valence band in the source and the conduction band in the channel. The model uses material parameters `me.tunnel` (tunnelling mass of electron) and `mh.tunnel` (tunnelling mass of hole) that can be defined in the material statements. The tunnelling current is exponentially dependent on the tunnelling mass and, therefore, `me.tunnel` and `mh.tunnel` should be used as parameters to calibrate the model with experimental results. For example:

```
material material=Silicon me.tunnel=0.14 region=2
models bbt.nonlocal qtunn.dir=1 bgn consrh conmob print
```

Here the electron effective tunnelling mass is defined to be 0.14. The model's statement defines the use of the non-local band-to-band tunnelling model (`bbt.nonlocal`) with the tunnelling direction *x* (`qtunn.dir=1`). The non-local tunnelling model works only with the `newton` method. Hence, we cannot use the `gummel` model together with the non-local tunnelling model.

We will now move on to the simulation of 3D TFETs with the example of a gate all around (GAA) nanowire TFET.

8.3 Gate all around nanowire TFET

In this section, we describe the simulation of a 3D TFET structure using ATLAS 3D. The most popular 3D structure studied is the nanowire transistor. Hence, we will study the simulation of a gate all around (GAA) nanowire TFET. The structure of the nanowire TFET is shown in Figure 8.7 The device is a p-channel TFET. The radius of the silicon nanowire (T_{Si}) is 10 nm and the gate oxide thickness (T_{ox}) is 2 nm. The length of the channel region is 200 nm and of the source/drain region is 50 nm. Hence, the total length of the silicon nanowire is 300 nm. The source/drain doping is 10^{21}/cm^3 and the gate work function is 4.8 eV. Table 8.2 gives the input file for the simulation of the GAA nanowire TFET.

The first and the most important step in any 3D simulation is defining a 3D mesh. Section 1 in Table 8.2 defines a 3D cylindrical mesh. The first statement (`mesh three.d cylindrical`) instructs the ATLAS to create a fully 3D cylindrical mesh. A cylindrical mesh is specified in terms of radius (r), angle (a) and z coordinate. In our nanowire TFET structure, the radius (r) is analogous to the y-direction of the 2D SOI TFET, the z-direction is analogous to the x-direction of the 2D SOI TFET and the angle (a) adds the third dimension. As done previously in the 2D SOI TFET (Section 8.1), the z-mesh should be finer at the source–channel junction and coarser at other positions. The r mesh should be finer at the silicon oxide interface and coarser at the centre. As shown in Figure 8.7, the 3D TFET has gate wrapped all around the structure and hence the electrostatics are invariant as we move along the angular (a) direction. Hence, we take a coarse mesh with equal spacing of 45° in the a-direction. In general, the mesh points in a 3D structure can be several orders larger in number compared to a corresponding 2D structure. Hence, it is necessary to define a coarser mesh for a 3D device structure as compared to a 2D device structure. A good practice is to start with a coarse mesh and then make it finer till the point where no change in

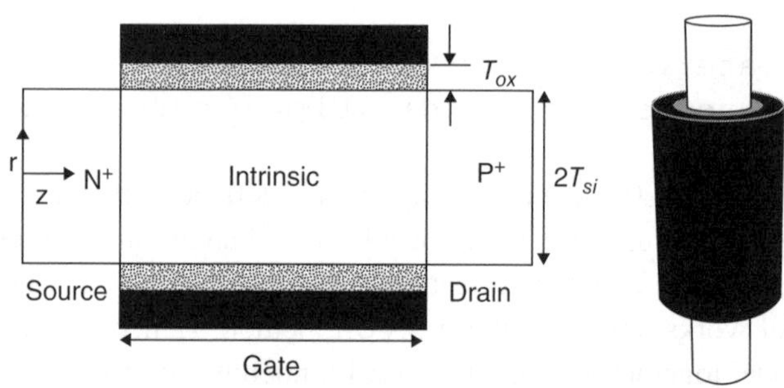

Figure 8.7 Schematic view of the GAA nanowire TFET.

Table 8.2 Example code in ATLAS for GAA TFET simulation

```
go atlas
TITLE GAA device simulation
#Section 1: Defining the 3D cylindrical mesh
mesh three.d cylindrical
R.MESH LOCATION=0.0 SPACING=0.002
R.MESH LOCATION=0.010 SPACING=0.0001
R.MESH LOCATION=0.012 SPACING=0.0005
R.MESH LOCATION=0.014 SPACING=0.001

A.MESH LOCATION=0 SPACING=45
A.MESH LOCATION=360 SPACING=45

Z.MESH LOCATION=-0.15 SPACING=0.025
Z.MESH LOCATION=-0.10 SPACING=0.002
Z.MESH LOCATION=0.0 SPACING=0.02
Z.MESH LOCATION=0.10 SPACING=0.002
Z.MESH LOCATION=0.15 SPACING=0.025

#Section 2: Defining the regions
REGION NUM=1 MATERIAL=silicon Z.MIN=-0.15 Z.MAX=0.15 A.
  MIN=0 A.MAX=360.0 R.MAX=0.010
REGION NUM=2 MATERIAL=oxide R.MIN=0.01 R.MAX=0.014 Z.
  MIN=-0.15 Z.MAX=0.15 A.MIN=0 A.MAX=360.0

#Section 3: Defining the electrodes
ELECTRODE NAME=source Z.MIN=-0.15 Z.MAX=-0.10 R.MIN=0.01
  R.MAX=0.011
ELECTRODE NAME=drain Z.MIN=0.10 Z.MAX=0.15 R.MIN=0.01 R.
  MAX=0.011
ELECTRODE NAME=gate Z.MIN=-0.10 Z.MAX=0.102 R.MIN=0.012

#Section 4: Defining the doping
doping  uniform p.type conc=1e20 reg=1 r.min=0 r.max=0.01
  a.min=0 a.max=360 z.min=0.10 z.max=0.15
doping  uniform n.type conc=1e16 reg=1 r.min=0 r.max=0.01
  a.min=0 a.max=360 z.min=-0.10 z.max=0.10
doping  uniform n.type conc=1e20 reg=1 r.min=0 r.max=0.01
  a.min=0 a.max=360 z.min=-0.15 z.max=-0.10
```

(*continued overleaf*)

Table 8.2 (*continued*)

```
#Section 5: Defining the material properties
contact    name=gate workfunction=4.8

#Section 6: Defining the models
models     conmob srh auger bgn fldmob print bbt.kane a.
  btbt=4e19 b.btbt=41e6 bbt.gamma=2

#Section 7: Solving the simulation
solve init
# do IDVG characteristic
method     gummel newton maxtrap=10
solve     vdrain=0
solve     vdrain=-0.05
solve     vdrain=-0.1
solve     vdrain=-0.2
solve     vdrain=-0.5
# ramp gate voltage
log      outf=test1.log
solve     vgate=0 vstep=-0.1 name=gate vfinal=-3
save      outf=test1.str
#Section 8: Reading the results
tonyplot   test1.log
tonyplot -3d nohw test.str
quit
```

simulation results are observed. The regions, electrodes and doping all have to be defined in cylindrical coordinates. In a 3D structure, all electrodes are required to have a finite thickness. Electrodes with zero thickness (i.e. 2D electrodes) will give a runtime error. To ensure convergence it is very important to include the Gummel method before the Newton method so that it makes a good initial guess of the solution. Without the Gummel method, the Newton method alone may take a long time to converge or may not converge at all. Also, if the simulations do not converge, it is advisable to take smaller bias steps and to define the `maxtrap` parameter, a value greater than its default value, which is 4. Maxtrap defines the maximum number of steps by which ATLAS will reduce the bias steps by itself (when the simulation is not converging) before exiting the simulation. Hence, by increasing the maxtrap value, ATLAS goes to smaller bias step values, which may make the simulation converge. The structure file of 3D simulations

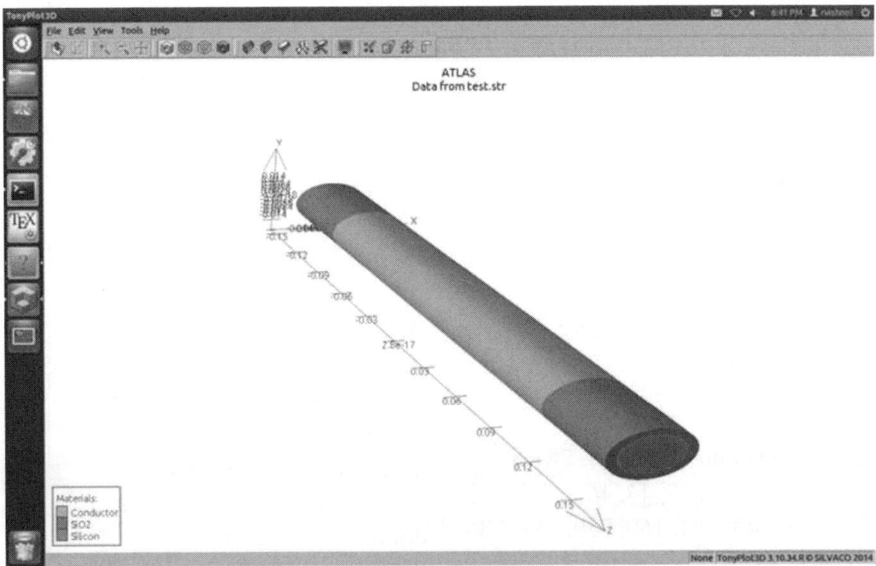

Figure 8.8 The structure file of the GAA nanowire TFET opened in TONYPLOT3D.

opens in TONYPLOT3D, which needs a graphics card above a certain configuration and may not open in all computers. The one in which it opens looks like that shown in Figure 8.8. To study the spatial variation in any physical quantity, we have to draw a cutplane in the structure, which will export all the data corresponding to that cutplane into a regular 2D TONYPLOT.

References

[1] *ATLAS Device Simulation Software*, Silvaco Int., Santa Clara, CA, USA, 2015.

[2] J. Wan, C. L. Royer, A. Zaslavsky and S. Cristoloveanu, "A Tunneling Field Effect Transistor Model Combining Interband Tunneling with Channel Transport", *J. Appl. Phys.*, vol. 110, no. 10, pp. 104503–104503-7, 2011.

[3] R. Vishnoi and M.J. Kumar, "Compact Analytical Model of Dual Material Gate Tunneling Field Effect Transistor Using Interband Tunneling and Channel Transport", *IEEE Trans. Electron Devices*, vol. 61, no. 6, pp.1936–1942, 2014.

[4] E. O. Kane, "Zener Tunneling in Semiconductors", *J. Phys. Chem. Solids*, vol. 12, no. 2, pp. 181–188, January 1960.

[5] A. Schenk, "Rigorous Theory and Simplified Model of the Band-to-Band Tunneling in Silicon", *Solid State Electronics*, vol. 36, pp. 19–34, 1993.

Index

Tunnel Field-Effect Transistors (TFET): Modelling and Simulation, First Edition. Jagadesh Kumar Mamidala, Rajat Vishnoi and Pratyush Pandey.